MARINE ELECTRICAL
CARE & REPAIR

By David MacLean

TAB BOOKS
BLUE RIDGE SUMMIT, PA. 17214

FIRST EDITION

FIRST PRINTING—NOVEMBER 1978

Printed in the United States of America

Library of Congress Cataloging in Publication Data

MacLean, David, 1921-
Marine electrical care & repair.

(Boatowner's how-to guides; 5)
Includes index.
1. Boats and boating—Electric equipment—Maintenance and repair. I. Title.
VM325.M3 623.85'03'028 78-11654
ISBN 0-8306-8946-X
ISBN 0-8306-7946-4 pbk.

Contents

Other TAB books by the author

BOATOWNER'S HOW-TO GUIDES:

Vol. 1—Hull Care and Repair
Vol. 2—Engine Maintenance & Repair
Vol. 3—Hauling Out and Winterizing
Vol. 4—Small Craft Electronic Equipment Care & Repair

INTRODUCTION

Of all the problems and failures that are to be experienced aboard small craft, those of an electrical nature will be first, foremost, and most persistent! Consider the following Coast Guard statistic: Most calls for assistance from small craft were the result of electrical failures—specifically, apparent or actual battery failure. In the previous statement the word *apparent* was included, because an after-the-fact examination revealed that often the problem was not battery failure but just a dirty, corroded electrical terminal. It is not difficult to conclude that some skippers are not yet practicing the principles of *planned preventive maintenance* (PPM)!

In all books of this series we take the position that small craft have enemies that attack the boat and its equipment, even when the boat has been snugged down for off-season storage. Among these enemies are elements that are usually thought of as beneficial, such as air, sunlight, and ordinary salt water. All are constantly at work to prevent things from working!

The primary power source of electrical energy aboard small craft today is the standard automotive 12 volt lead-acid storage battery. This battery is referred to as a surface charge battery, meaning that its designed purpose is to start some type of internal combustion engine, at which point the alternator or generator assumes the electrical load. Only the surface charge of the battery is used to start the engine, and the charge is immediately restored by the action of the alternator/generator. For automotive equipment, such a system is satisfactory; for small craft it is not, because in many instances the electrical load continues long after the engine is shut down. For example, while the boat is at anchor, the depth sounder/fish finder may be used, the radio telephone may be left on to monitor the emergency channel, the anchor lights may be on, and the entertainment radio may be playing, all of which quickly peels off the surface charge of the battery and leaves little to start a modern high compression engine. The capacity of an automotive surface charge battery used aboard small craft is marginal, and it is absolutely necessary that the battery be reasonably fresh, carefully maintained, and constantly watched for its state of charge. A number of solutions to this problem will be discussed and described in detail in the following chapters of this book.

The enemies that attack electrical equipment on small craft are often subtle and are rarely dramatically evident. For this reason planned preventive maintenance may be the only way to anticipate and thus avoid electrical failures. It will be helpful if the small craft skipper is acquainted with the nature of the attacks on electrical equipment. First, in order to be functional, electrical equipment and its associated wiring must use *active* metals, that is, metals that conduct electricity. For example, copper is an active metal; stainless steel is not. Aluminum is an active metal; silicon bronze is not. Active metals easily form undesirable chemical combinations in the presence of oxygen, salt, and heat. The green-grey patina that quickly forms on copper wire and terminal lugs is copper oxide and is, for the most part, a poor conductor of electrical current. A similar substance forms on the surface of aluminum.

As if this were not enough, there is often a galvanic action between electrical conductors and the associated metals of the equipment. Galvanic action is electrochemical in nature and is the result of placing two dissimilar metals in close proximity to each other separated by an electrochemical substance such as salt water or a mixture of acid and water. Under these conditions an undesirable electric current will flow from the more active of the metals to the other. When this action is allowed to continue, there will be a gradual eating away and corrosion and eventual failure of the active metal.

Added to the problems of surface corrosion from air contact and galvanic action are heat problems. For example, a terminal lug is attached to the end of a wire carrying an electric current. If either the lug or the point to which it is attached becomes corroded at the contact interface, contact resistance results. Electric current flowing in the presence of this resistance will generate heat, and heat causes metals to expand. Different metals expand and, as they cool, contract at different rates, and this action hastens poor electrical contacts and causes electrical failures both at the contact point and further downstream within the equipment. If the resistance is great enough and the current high enough, a fire may start.

Add vibration to this list of enemies and it is small wonder that the small craft electrical system functions at all! The vibration from the operation of the engine, the normal rolling and pitching of the boat, and slamming through the water at high speeds will separate any loose electrical contacts and cause an electrical failure that must be attended to on the spot—and that spot is rarely convenient!

For the most part these enemies cannot be eliminated, but they can be controlled. A well devised and faithfully carried out planned preventive maintenance program and the use of modern materials that have been developed for marine service may well be the only answer for the small craft skipper who seeks to prevent and correct small craft electrical problems.

What might be thought of as the crown of small craft electrical maintenance problems is the lack of available professional advice and service. A few of the largest full service yacht yards might offer an electrical shop staffed by skilled practitioners, but their services are generally limited to the larger yachts, and the small craft owner must turn elsewhere. Some engine mechanics will take on an electrical problem, but it is not their basic forte and they charge correspondingly. Licensed electronic technicians in the radio shops can do the work but must charge for their time as if the job were electronic in nature. Perhaps in the future there will be regular electric service shops for small craft staffed by skilled electricians. The Coast Guard and a number of boating organizations have long recognized that electrical problems on small craft exist and suffer from lack of cures. New safety standards now in effect for boat manufacturers should result in better quality electrical systems than those designed for automotive use and adapted for marine use.

In the meantime the small craft skipper must solve his electrical problems himself, and that is what this book is all about. Our concept of planned preventive maintenance has as its goal the prevention of problems by recognizing them *before* they occur. Leaders in business, industry, and particularly the military subscribe to the PPM concept, and millions of dollars are saved that would otherwise be spent on *corrective* maintenance, which requires expert labor, costly parts, and (perhaps worst of all, at least to the small craft skipper) the lack of availability of the unit during the time it is in the repair shop.

On the other hand, small craft preventive maintenance seldom requires skilled labor, exotic tools, and a great many parts. For the average skipper all that is required is the desire to keep his boat in top condition and a little guidance on what to do and how to do it. A boat on which a PPM program has been faithfully carried out is safer, better looking, has greater resale value, and perhaps best of all is more available for use because it runs better.

1 PLANNED PREVENTIVE MAINTENANCE

If you have read any of the previous books in this series you will recognize that we make every effort to persuade you that planned preventive maintenance will work for you! Our method is quite simple and does not present any new and startling revelations. It is true that getting your own PPM program started involves a bit of research and analysis; but once these tasks are complete, planned maintenance becomes an easily followed routine that offers great possibilities for a handsome return in maintenance profit.

PPM is based upon the premise that most or nearly all of those tasks that can be described as preventive maintenance can be done by any small craft skipper, no matter what his shoreside occupation, and applies to all watercraft, both large and small. By contrast, *corrective* maintenance in almost all cases requires the expensive services of professionals, the use of exotic tools and materials, costly haul-outs, and long delays to obtain parts suitable for marine application. Our concept can be expressed by paraphrasing Murphy's Law (our feeble attempt at wry humor): Whatever there is in, on, or around a small craft that *might* go wrong—it will! Knowing that this law is applicable, you can do much to offset its effects. When you know the nature and effects of a small craft's enemies, you can begin to formulate a strategy to bring them to near total defeat!

We have observed a number of highly motivated and industrious skippers who seem to have maintenance programs, but who nevertheless often run from one maintenance brush fire to another, never quite catching up and never fully enjoying their boats. It's almost as if operating their boats would interfere with their maintenance programs! These skippers suffer from the lack of an organized and systematic approach to preventive maintenance.

We would be the last to contend that preventive maintenance is as much fun as cruising, fishing, water skiing, or just plain loafing in the boat. Yet many of the onerous chores can be lightened and many of the repetitious tasks can be substantially reduced to the point where boat use time is actually increased by systematic planning and the application of that planning. There is a lot of confused error associated with making a list of things to be done, either to get the boat ready for the season or just to keep it going. Often when such a list is composed and studied by the skipper he becomes so discouraged by the number of tasks that he tends to sluff over them and resorts to the most rudimentary patch and glue cures, hoping against hope that they will get him through another season. The next time around the list gets longer and more complex—often to the point where the boat is put up for sale at a potential loss to the skipper. This vicious circle can be broken, for every skipper *does* have the skill and resources to do preventive maintenance and he *does* have the time and resources *if* he will but formulate a simple and personal plan and stick to it. The plan must be viable and allow for modification and change to suit the skipper's needs, temperment, and resources.

Any plan, in order to be minimally effective, must be based upon *hard data*—factual information stripped of all opinion. This data can be gathered by the small craft skipper from such sources as a careful inspection program of the various systems that make up a small craft. He must know what to look for—the effects of the enemies' attacks. There is historical information that can be gathered from logs, records, and equipment manuals. Thus the first step in formulating a personal preventive maintenance plan is to gather all the hard data available.

You will find as you begin to collect this information that there is no lack of data and often seemingly too much. You will then have to winnow the wheat from the chaff, keeping the useful and discarding the useless. In this book we will restrict our data-gathering suggestions to the electrical systems and equipments aboard small craft.

Data Gathering

For the average small craft skipper there is no better method of gathering useful maintenance data than the old-fashioned "eye-

Fig. 1.1 No. 1 troublemaker. Note layer of non-conducting "crud" that has formed on the surface of the battery terminal.

ball" technique. However, the advanced symptoms of trouble brewing in electrical systems are often quite subtle. As mentioned, the greatest single cause of calls for Coast Guard assistance is *apparent* battery failure that is later diagnosed as corroded terminals or another problem that could have been cured easily by the skipper—had he bothered to look!

The large main terminals of lead-acid batteries, both positive and negative, and their associated cable clamps are fabricated of active (electricity-conducting) metals, and, once installed in the boat, they are exposed to subtle but effective attacks by the oxygen and salt in the air and the ever-present moisture. Compounding the problem are battery gasses, which are emitted through vents in the filler caps as the battery is charging and form undesirable chemicals on the surfaces of the battery terminals and cable clamps. That these chemicals, which appear as greenish-grey crystalline substances, are poor electrical conductors goes almost without saying. Sometimes these chemical deposits are evident on the outside of the terminals, but more often they are not. The worst and most subtle problem is that they will also form *inside* the cable clamp and the interface where the clamp attaches to the terminal post. Chemical deposits inside the clamp will cause one or more of the following effects: As the starter key is operated, the starter

Fig. 1.2 Inner surface of battery cable clamp also must be painstakenly cleaned before reconnecting.

motor fails to turn the engine over. A click or often a chattering of the starter relay may be heard. Lights that are on anywhere on the boat will sharply drop in brilliance, which is a good clue to the problem.

And now the point of this admittedly tedious explanation. The PPM skipper, knowing that battery terminals suffer from terminal illness, will routinely and on his own, schedule loosen and remove the cable clamps from the battery, inspect both the insides and outsides, and burnish with sandpaper or a special tool both the inside of the cable clamp and the outside of the battery terminal. He will then rebolt the cable terminal in place and lightly coat the terminal with either grease or petroleum jelly. The entire task requires less than ten minutes and only the most basic of tools and materials—things normally kept in the boat box. Thus, the PPM skipper *prevents* the problem from happening at all!

Throughout the boat much the same things occurs. The unfriendly air, heat, and vibrations attack the hundreds of electrical connections found on the smallest of small craft. However, only the most crucial connections need to be disassembled and inspected, such as those at the battery and at the engine. For example, the ground connection on the engine block is rarely inspected, cleaned and tightened. There are a number of con-

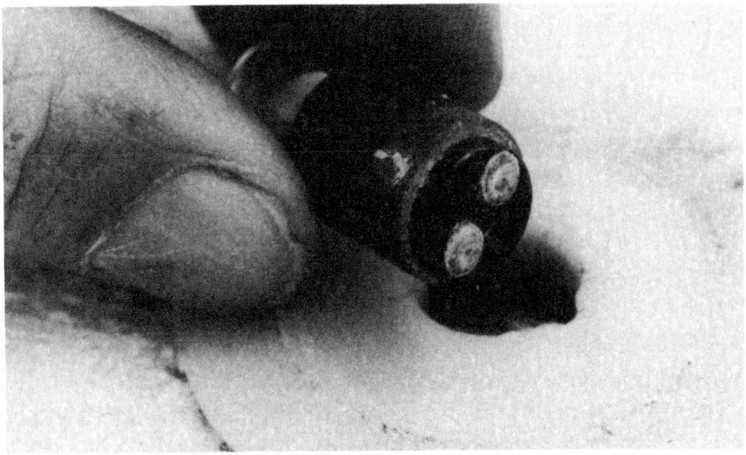

Fig. 1.3 Prior to cleaning the dual terminals, the lamp burned weakly. Cleaning, burnishing, and WD-40 spray restored lamp to full brilliance.

nections usually found on the starter solenoid that act as a central distribution point. The connecting cable terminals and studs of any and all equipment that depends on a heavy supply of current in order to operate properly should be inspected periodically.

Lighting problems may also be the result of poor connections. What initially seems to be a burned out light bulb may be no more than a badly corroded socket and terminal. Almost all terminals can be tested for tightness by simply attempting to wiggle the connection with either your fingers or a pair of needle-nosed pliers. Terminal lug screws will loosen for no apparent reason (actually it is vibration) and cause a contact problem.

The initial inspection plan should include all installed electrical equipment as well as the primary power supply (the battery or batteries and the alternator/generator) and all of the main power connections. If a motor is involved, try to get a look at the brushes and commutators. Check out all lighting circuits, but *not* by simply switching the lights on and off! Disassemble glass and plastic covers and give the bulb and its socket a good going over. Carefully check fuse and distribution panels, paying particular attention to all grounding wires. Inventory your spare electrical parts, wire, fuses, tape and any special tools needed to effect repairs and maintenance.

If in your preliminary inspection you keep in mind the nature of the electrical problems that most often beset small craft skippers, then you can prevent most if not all of them. Most troubles will stem from one or more of the following: electric current leakage due to faulty insulation or damp or dirty surfaces; abnormal wire resistance caused by corroded conductors; broken wire strands due to chafing; bad contacts due to loose and/or corroded connections; breaks in the wire due to unnoticed stresses; and, most dangerous of all, simple short circuits.

Current leakage

Current leakage will put a constant drain on the battery and can cause corrosion, especially when the boat is left unattended during the week. If when the weekend rolls around you try to start the engine only to find the battery seemingly dead, current leakage may be the cause. To determine if current drain is the problem, turn all electric switches off, remove the positive battery terminal cable clamp, and connect a small low current test lamp between the battery terminal and the cable terminal. If there is the slightest visible glow from the test lamp filament, you have a definite current leakage problem.

The next step is to locate and identify the circuit that is causing the leak. If you have a central distribution fuse panel, it is a simple matter to pull the fuses one by one while observing the lamp. If the lamp should suddenly go out as you disconnect a fuse, you have pinpointed the guilty circuit and can inspect the wiring of that circuit for chafed spots or any point where the positive wires may be making a partial contact to the negative return. Inspect also for wet lamp sockets and partially worn insulated wires.

Another but rarely suspected source of current leakage is from the top of the battery, where dirt, moisture, and battery gases combine to form conducting paths along the surface of the battery between the positive and negative terminals. To cure, carefully wash the surface of the battery with baking soda and water, taking pains to keep this washing solution well clear of the filler caps and vents. Rinse with clear water and carefully dry the top of the battery. A number of insulating sprays are available that are effective to a limited extent in preventing this problem.

Another method of keeping tabs on battery leakage is to carefully observe the position of the ampmeter (if your boat has

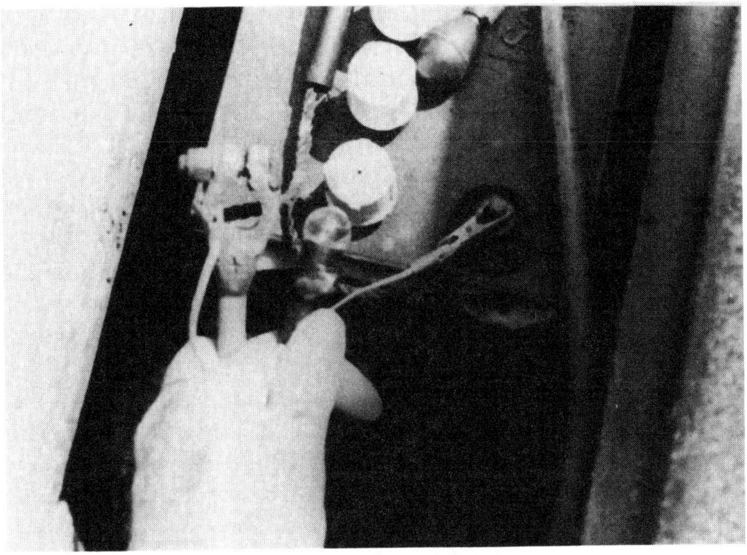

Fig. 1.4 Current leakage test with 12 volt test lamp. Lamp is connected between the battery terminal and the cable clamp that has been removed.

Fig. 1.5 Prevent other leakage paths by giving the top of the battery a good bath.

one) when the engine and all electrical circuits are off. The needle should *never* read discharge or negative at this condition. It should be understood, however, that the ammeter is designed to indicate rather large amounts of current and may not always clearly indicate a small leakage current.

In your inspection be suspicious of any and all wire splices and the crimp-type, so-called solderless terminals, which tend to corrode inside of the lug sleeve. On small craft you will often find jury-rigged splices to equipment that was added to the boat. Air, dampness, and salt tend to penetrate the point where the wire and lug are supposed to make clean electrical contact and form various oxides that inevitably end up creating a high resistance contact point. When that circuit is energized, the current through such a point creates a voltage drop and begins to get hot. The equipment runs less efficiently or not at all, and the danger of fire is increased. No terminal or wire on the boat should ever feel even warm to the touch. Lamps and sockets are of course excluded from that statement.

Once you have completed your inspection of the boat's electrical system(s), you should sit down and study your list of discovered faults. Rarely will you be able to do a thorough job of inspection without finding something that requires maintenance attention. Include in your analysis any logs you might have kept, and be sure to check through engine and equipment maintenance manuals for added information and periodic preventive maintenance procedures.

ANALYSIS OF DATA

Your analysis of the results of the electrical system inspection and other data must have as its objective the assignment of priorities to the necessary tasks and the determination of which tasks you can do yourself and which require special or outside assistance. With the tasks reordered and the priorities set you can begin to draft your own *personal* preventive maintenance plan. Once you have corrected any major problems, your PPM plan will often amount to little more than periodic and scheduled inspections to be done most during the operating season, for there will be other times when routine painting, refurbishing, and polishing are called for.

As an integral part of your plan, try the technique of thinking through each task before it is actually begun. If you jot down the steps required you will be able to list and gather in advance all the tools, parts, and materials that you will need.

As a final step in drafting your PPM program, be sure to include a schedule for the completion of each needed task, taking care to allow a little more time to accomplish the job than you really think you need. Once you begin the job, work at a steady but leisurely pace. Try not to set impossible and hence demotivating goals. At the end of a given job try to quit on a peak with one good job well done. Don't start a second job late in the day and have to stop before dark. Get yourself psyched up to do the work and stick to your schedule. Before you know it you'll be enjoying a well earned rest while running a boat that should be reasonably free of at least electrical problems.

THE NATURE OF ELECTRIC POWER

A basic and practical idea of the forces that you'll be working with will help you to carry out most preventive and corrective electrical maintenance tasks. However, we are not about to try to make an electrical engineer out of you in a few paragraphs, so relax and skim through this section.

Almost all small craft today are equipped with 12 volt DC electrical systems. Some of the larger yachts have 24 or 32 volt systems, and some of the yet larger boats have auxiliary diesel generators that make available 110 and 220 volt AC power. Since we are mainly concerned with the ubiquitous 12 volt systems, we'll stay with those.

A *volt* is a unit of electrical pressure. It is what pumps the current through the circuits. It is common practice to think of volts in much the same way as you think of the water pressure at the outlet or discharge side of a water pump. Water pressure is most often measured in pounds per inch. And it is this pressure, much the same as volts, that provides the push to make things work. Once you get into electronics you will find that volts are further divided into millivolts (one-thousandths of a volt) and, smaller yet, microvolts (one-millionths of a volt), which is indeed a very small amount. Voltage also is often referred to as electromotive force (EMF) or potential, terms that are used—often to confuse—interchangeably. In this book we will try to stay with plain old volts!

In an electric circuit the stuff that moves and does all the work is *current*. To this day there is still a lot of confusion and controversy as to just what current is. The best cop-out definition we've ever heard is: current is the net movement of electric charges in a conducting circuit. Note that the definition avoids the trap of saying what an electric charge is; and it does not get into the pitfall of the direction of current flow. It is convenient to think of current flow as similar to water in a pipe, and in many respects their behavior is often the same.

The unit for current flow is the *ampere*, which is often abbreviated to amps. Like voltage it may be seen more often in smaller amounts such as milliamps (one-thousandths of an amp) and smaller yet microamps (one-millionths of an amp). Always keep in mind that it is current flow that kills, not voltage. When both current and voltage are present in sufficient amounts to overcome the natural resistance of the human body, a dangerous and sometimes fatal electric shock can occur. The electric symbol for direct current flow in amperes is I. For voltage it is V or more often E.

In addition to current and voltage there is always present in an electric circuit some amount of *resistance*, which may be defined as that property of an electric circuit that tends to oppose the flow of electric current. It is easy to think of resistance as the friction inside a pipe or hose when water is flowing through it. The unit of resistance is the *ohm*, and the symbol most often used is R. Often much confusion is associated with the unit of resistance. For example, in an open circuit the resistance is said to be infinite or maximum since current cannot flow. In a short circuit, resistance is, for all practical purposes, zero, since the current that flows is limited only by the pressure and the current capacity of the source. Fortunately for we small craft owners there are simple, reasonably accurate, and inexpensive meters that will measure each of these values and that often can be used to locate and identify the source of an electrical problem.

The last unit that the average small craft skipper should have any concern about is *power*, defined as the *rate* at which work is done. The unit is the *watt* and the electric symbol is P. For example, most electric lamps are rated in watts, as are toasters, electric irons, and stoves. It is these watts, normally expressed in thousands of watts or kilowatts, that you are billed for each month from the power company. You will find that on a small craft, because of the low voltage 12 VDC system, very few things will be rated in terms of watts. However, if you are curious to know how many watts any device aboard is using, you have but to multiply the voltage (almost always 12 volts) times the current that the device normally draws. In a similar manner, if you know the watt rating of a lamp and its required voltage, you can easily determine the current that it will draw from the battery. Simply divide the voltage (*E*) by the power (*W*) and the result is in amperes (*I*). Thus a 6 watt running lamp will draw 2 amperes from the battery all the time it is left on. Small wonder the battery is quickly run down and will not start the engine when lights are needlessly left on!

SOME PRACTICAL APPLICATION

Conductors and Insulators

From the previous paragraphs on voltage, current, and resistance we now move to the practical application of that little amount of electrical theory. In order to do work a pump (your 12 volt battery, for example) must be connected to a load (a 12 volt bilge pump, again for example) by means of the proper size (diameter) of electrical conductors or wires. It is essential to understand that there is no such thing as a perfect conductor. There are metals that size for size will conduct a greater amount of electric current without raising the temperature of the wire than others. Unfortunately, these are not practical, for they would have to be either platinum (the best) or pure silver (the next best). These exotic metals are followed by copper and then aluminum in the table of good electrical conductors. However, if you have had the need to purchase any amount of copper wire or tubing lately, you know that the price of copper is rapidly catching up to silver, which may well be an excellent reason for reducing the number of wiring mistakes on small craft!

The essential difference between a good conductor and an equally good insulator is found in the number of free electrons present in a given substance. The term *free* here means that the electrons, which are normally closely attached to their parent atoms, are free to move from atom to atom when an appropriate voltage pressure is applied. These free electrons tend to avoid or to be repelled from a point where electrons are said to be in excess (such as the negative terminal of the battery) and flow toward a point where the supply of electrons are said to be deficient (the positive terminal of the batter). This theory of electron flow (the net movement of electric charges) in a conductor is to this day considered sheer hearsay by old-time electricians. The more conventional theory of electric current flow holds that something (they never say what) flows from the positive terminal of the battery to the negative. For a small craft skipper intent on maintaining his boat's electrical system it is sufficient that both theories work equally as well. This is not to imply that polarity in small craft is not important, for it is; and in some circumstances it is desperately so, as we shall explain in later chapters.

If you will again think of current flow in conducting wire as similar to water flow in a pipe or hose, you will readily see that a larger diameter of pipe will carry a greater volume of water for a given pressure per unit of time. Thus the greater the diameter of the wire (or better still the cross-sectional area) the greater amount of current it will safely carry. It is with regard to safety that the analogy between water and electric current no longer applies.

All good conductors, even silver and platinum, have a certain amount of resistance to the flow of an electric current. As the current in any given wire is increased, the resistance remains the same. Current through a resistance generates heat. The more current, the more heat, until a point is reached where it is no longer safe and might, in extreme circumstances, even melt the wire and cause a fire.

Thus it can be readily seen that the thin little wires used to connect a cabin light bulb cannot be used to carry the hundreds of amperes of current required by the starting motor on the engine. It is for this reason that various boating organizations and the Coast Guard have established safety standards for small craft wiring. Even though you are not a boat builder you must observe these safety standards if you are running a new set of wires in your boat or making repairs to any that are already installed. Common sense will tell you, for example, that you should never splice a smaller wire into a larger diameter wire, no matter what the reason. Using a wire size that is too large for the current it must carry will rarely hurt (except in the wallet); while too small a wire invites disaster and may send both you and your boat to that big marina in the sky.

This safety standard is applied to the capacity—or as it is technically termed *ampacity*—of wire in its ability to carry a given amount of current for a specific size in terms of a limited rise in temperature. This concept is difficult for the average skipper to understand and apply. To make matters easier, a given size of wire should carry its rated amount of current without a percentage of voltage drop. And again the water in the pipe analogy becomes helpful to understand this restriction. As water flows down a pipe the friction and the constraints of the size of pipe will effect a decrease in pressure along the length of the pipe. The same idea holds true with an electrical conductor. Most electrical appliances function best when fed with the proper level of voltage for which they were designed. Any losses in the feed wires means a loss of operating voltage at the input terminals of the device. Thus a 12 volt lamp glows much dimmer if the voltage applied to the base of the lamp is but 9 volts and the remaining 3 volts are lost in the wire that is carrying the current and voltage to the lamp. However, to keep costs and other wiring problems down, we allow a 10 percent voltage drop in non-critical things such as cabin lights and other devices not involving safety or operation. Voltage-sensitive devices such as navigation lights, electronic equipment and various DC motors should be fed with wires in which a 3 percent voltage drop is not exceeded.

Insulation

When there is enough voltage present, a current flow will take place in many things where it is not desired. A film of moisture and dirt on the top of a battery will often conduct enough leakage current to run down a battery to the point where it will not start the engine until the battery has been recharged ashore. The same applies to wire conductors. To restrain and confine the current and voltage to only those areas where it is desired, wire is covered with various forms of insulation. Rubber, neoprene rubber, and various forms of thermoplastics are used as insulating materials to cover and protect wire. Since almost all voltage on small craft is 12 volts of DC, only a thin layer would be thought to be sufficient. However, it is common practice to use wire with an insulation resistance that will stand up to 600 volts. The heavier insulation resists cracking and deterioration from aging, heat, sunlight, gasoline, oils, and greases. Different insulating materials are needed for different applications on the boat. A wire with thermoplastic insulation would be unsuitable for use around the engine where it might be attacked by the heat. The table in Chapter 6 lists the application of the wire, the designation by a three-letter code, the maximum temperature in Fahrenheit in which it is to be used, and a general description of the wire.

Suppose, for example, that you have decided to install a voltmeter at the operating position on your boat so that you can monitor the behavior of the alternator and the condition of the battery—an excellent idea by the way. Voltmeters for marine application do not draw a significant amount of current, so you would choose a rather light size of wire such as a no. 16 (AWG). However, since the pick-up point to connect the panel-mounted meter is in the engine compartment, you must select a type of insulation such as THW or RHW.

Color Codes for Wire

The small craft skipper is often appalled when he first views the back side of the control and wiring panel on his stock boat. Often, running down a short circuit or a high resistance connection is made difficult by the production wiring, the only objective of which seems to be to get everything connected in the quickest, least sensible manner. Yet many of the stock boat manufacturers are now subscribing to a standardized color coding system promulgated by the American Boat and Yacht Council. Thus, in addition to the best and safest size and material used as insulation, each wire will have a color that indicates its application and the type of circuit it is associated with. Some wires are a solid color and others will use a second color in a stripe or as it is called a "tracer." If you were to trace out the ammeter circuit in order to locate and eliminate a poor connection, for example, you would look for an orange, rather heavy wire and find that it is connected to the ammeter terminals at the instrument panel and to the alternator in the engine area.

MATERIALS AND TECHNIQUES

In the following discussion we will list and describe many of the devices and materials that you will find helpful in any PPM program. All of these materials should be available in marine hardware stores, electrical and electronic supply stores, or automotive supply stores, so that the average small craft skipper has immediate access to them.

Crimp-type terminals For almost any electrical application (other than boats) the solderless, crimp-type terminal is considered sound and safe electrical practice. Its convenience cannot

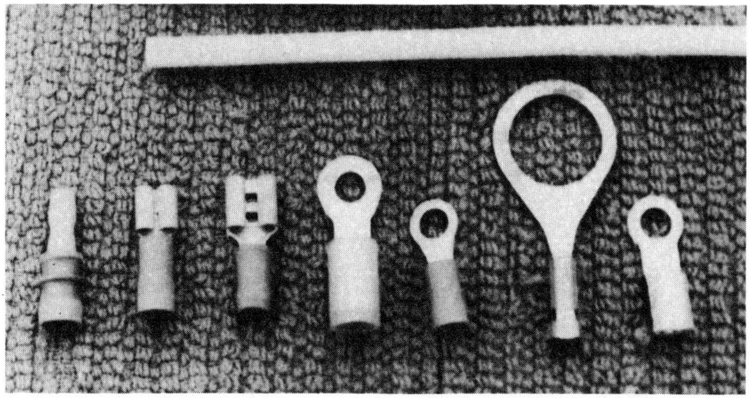

Fig. 1.6 Examples of crimp lugs for small craft wiring jobs.

be denied. Using crimp terminals and an inexpensive tool, you can quickly and effectively make a connecting wire that will serve well. On small craft, however, which are exposed to excessive moisture and salt, the crimp terminal may be more of a troublemaker than an advantage. More often than not the interface between the sleeve of the terminal and the strands of the copper wire will, over a short period of time, acquire a patina or coating of a greenish-grey powdery substance that is not a good electrical conductor. This substance adds resistance to a critical and undesirable point. While the mechanical part of the connection will still be good the electrical part will deteriorate to the point where a severe voltage drop will appear between the end of the wire and the terminal lug sleeve. To make matters worse, inspection will rarely reveal this problem because it is hidden under the sleeve and the insulation. Nor is there any easy way to test for the problem. Only cutting the sleeve of the terminal with heavy nippers and pulling it apart will reveal the coating that has formed. Experience has shown that the cure is simple and involves little more than soldering the "solderless" crimp-type connectors.

Crimp-type connectors come in a wide variety of shapes, sizes, and applications. The ring-type is preferred for small craft and is selected for a sleeve size suitable for the size of wire it is to be attached to and the size of ring that is to slide over the terminal stud. Thus a no. 10 crimp terminal will fit well over a no. 10 stud and accept either a AWG size 14 or 12 stranded wire.

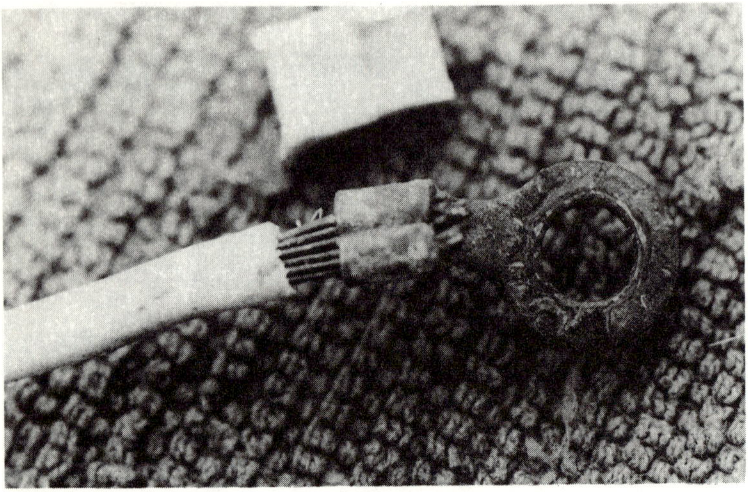

Fig. 1.7 Insulated cover has been cut away to show the formation of non-conducting corrosion on both the lug and the strands of wire.

Further safe and good practice requires that no more than four conductors should ever be attached to one terminal stud and that the connections be made so that there is no strain on the stud.

When attaching the crimp lug to a stud terminal use the proper hardware, which includes a lock washer to prevent the terminal from loosening under heat and vibration.

Crimp terminals should *never* be attached to a wire using any other tool than the special crimping tool. If you try it with ordinary pliers you'll crush the sleeve and the terminals will not last.

The following step-by-step procedures assure a good, trouble-free, wire terminal connection. The only preventive maintenance ever required will be an occasional removal of the terminal from the stud and cleaning with sandpaper to burnish the contact surfaces.

To attach a crimp type lug:

1. Select the correct size of crimp terminal for the wire that you are using and for the size stud that the ring of the terminal must fit.

2. On most crimping tools there are different sizes of wire stripping cutting holes, which are numbered to correspond to the

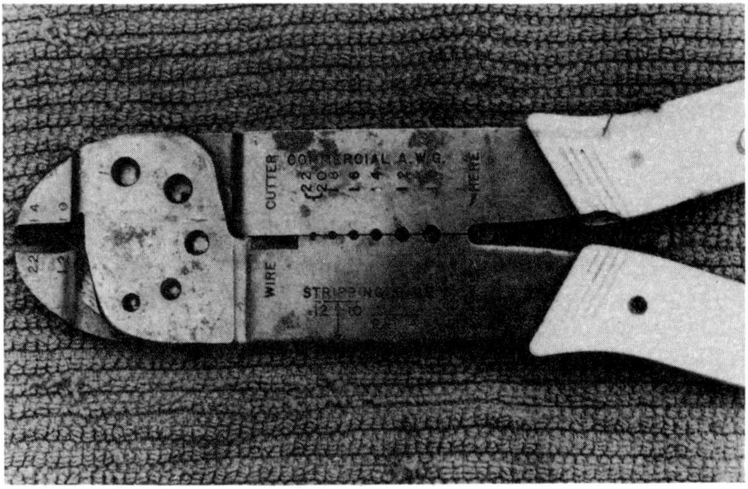

Fig. 1.8 Insulation stripping knives on crimping tool for wire sizes 10 to 22 AWG.

size of wire they will strip without cutting or nicking the soft strands of copper conductors. Thus you use a no. 10 hole to strip a no. 10 wire. Insert the wire into the proper hole and close the tool. Rotate the tool back and forth to complete the cutting action on the insulation and pull the wire free from the tool. The insulation will drop free. Strip only enough insulation to allow the wire to just barely project out of the sleeve with the insulation snug against the sleeve.

3. Carefully select the correct size and type of crimp lug from your collection in the boat box. If the lug is to be used to attach a wire to a screw terminal, then select a ring-type lug that just fits over the terminal stud. The sleeve of the crimp lug should be of a diameter that will accept all of the strands of wire on the stripped end once they have been twisted together with a rolling action between the thumb and finger.

4. Insert the stripped and twisted end of the wire into the sleeve of the crimp lug, pressing the insulation of the wire snugly against the base of the crimping sleeve. Note the amount of excess wire that protrudes out of the sleeve, if any, and mark to cut off the excess. It is best to allow about 1/8 inch of wire strands to protrude from the sleeve for other steps.

5. Your crimping tool may have more than one mandrel for the crimping to allow for different sizes of lugs (sleeves). Set the lug sleeve into the appropriate sized mandrel, making sure that the wire is snug in the sleeve with no bare wire showing at the wire end of the sleeve. Operate the crimping tool to the closed position until the tool's stops are touching. If the crimp lug has an insulated sleeve, the crimping action is final.

NOTE: If you opt for a better and less troublesome connection using an uninsulated type of crimp lug, you should obtain some of the new shrink tubing to use for the insulation. This tubing comes in various diameters, and a small amount of various sizes can be easily kept in the boat box for future needs. Cut a small section of shrink tubing, enough to cover at least 1/8 inch of the insulation of the wire and the entire metal sleeve of the crimp lug. Slide the tubing onto the wire *before* you operate the crimping tool. Once the lug has been crimped to the wire the tubing can be slid up the wire into position to cover the wire and lug sleeve. Apply a modest amount of heat to the tubing either by using a match or by holding a soldering iron tip close to but not touching the tubing. As heat is applied the tubing shrinks smaller and smaller forming an insulated and airtight covering over the sleeve of the lug and the wire where it enters the lug. Shrink tubing can be found in most electronic parts and electrical supply stores. If you have trouble finding shrink tubing it should be noted that plastic electricians tape applied to the sleeve of the lug and wire while the sleeve is still hot from the soldering iron will behave much like shrink tubing and bond tightly to the sleeve and the insulation of the wire.

6. The final step to make a truly trouble-free connection is to apply heat to the ring section of the lug at the base where the protruding end of the stripped wire can be seen. Heat may be applied using a small soldering iron if electric power is available or one of the new and very small gas torches. Use only enough heat so that when the wire is touched with rosin core (never acid-type) solder, the solder will melt and run into the strands of the wire and the base of the lug ring. A little dab of solder is usually more than enough. Too much and the ring may not fit over the terminal stud.

Basically what this technique aims to accomplish is a tight mechanical and electrical contact completely devoid of any

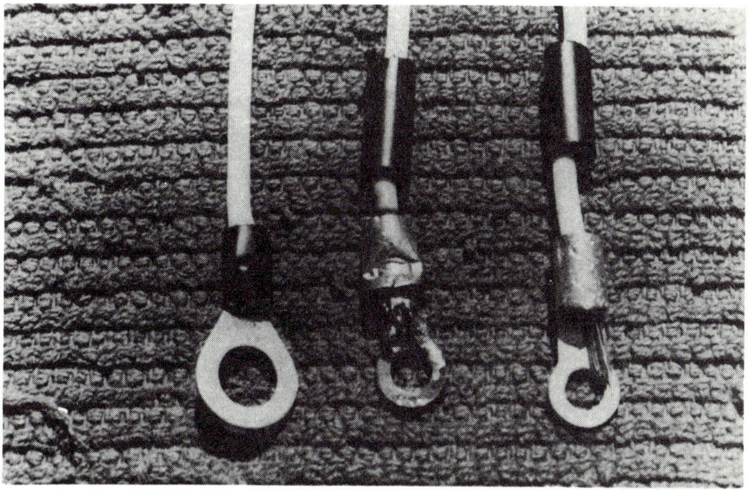

Fig. 1.9 Step for making a marine service connection: Wire is stripped and shrink tubing slid onto wire with lug; lug is crimped and soldered with rosin core solder; and light heat is applied to shrink tubing to make a trouble-free connection.

contact resistance between the crimp lug and the wire. In addition, the shrink tubing (or plastic electricians tape) together with the touch of solder prevents air and moisture from entering the sleeve of the crimp lug and stripped wire. The solder is essentially for the same purpose—to provide a strong connection and keep the air and moisture out. Although oxides, and therefore contact resistance, can still form on the ring section of the lug, these can be easily seen and prevented by from time to time burnishing the rings clean with a bit of crocus cloth or sandpaper and then spraying with WD40 or a similar insulating spray.

Splices

Generally, splices aboard small craft are an abomination. Yet it is recognized that there are always times where a splice must be made. A good example might be the need to install a plastic case bilge pump. Most of these pumps come with molded-in wire leads and no other terminals. Feed wires coming from the switch panel and negative terminal of the battery must be spliced to the leads of the pump to complete the installation. More often than not the splice will have to dwell in the nether regions of the bilge area along with the pump and thus be exposed to moisture, heat, corrosive fumes, and vibration. If there is even a small amount of contact resistance at the splice points, the pump will not get its required full voltage and current, the pump motor will not run at full speed, and the pump will not do its job adequately.

Nearly all of the electrical appliances and equipment aboard small craft are voltage sensitive; that is, they are designed to run at maximum efficiency at 12 volts DC and nothing less. To compound the problem, only 12 volts are available! Poor connections will, therefore, consume an often significant portion of the 12 volts, leaving the gear to do the best it can with what is left.

There are several crimp-type devices that can be used to form reasonably reliable splices, and these should be attached to the feed wires as suggested in the previous series of steps. When considering the use of crimp-style splicing devices, keep in mind that if any metal is left exposed *after* the splice has been made you will eventually have trouble at that point. To prevent future problems, carefully spray and cover the exposed metal of the splice with shrink tubing or plastic tape.

Often there is a practical requirement for quick-disconnect types of splices so that equipment may be removed from the boat either for security or for service. Again there are special plug-type splicers that are arranged so that the polarity of the connections will not go astray. When selecting a splicer to use on the boat, try to think of all the things that might go wrong at the splice point if you use that device—then make your choice.

When a splice must be made and a crimp lug splicer is not available, the best choice is the Western Union splice. Approximately 1 to 2 inches of insulation is stripped from each of the ends of the wire to be joined. The end of one wire is twisted in a clockwise direction tightly around the shank of the other wire. The opposite end is then twisted just as tightly in a counterclockwise direction on the other shank. Any excess wire is trimmed with diagonal cutters and heat and rosin core solder are applied to the entire splice area. The bare wire may be covered with shrink tubing or tightly wound plastic electricians tape. Taking a small amount of extra time and care to make your splices airtight, moistureproof, and mechanically and electrically sound will pay off in a trouble-free connection.

Wire Clamps, Cable Straps, Feed-Through Grommets

Chafe is an enemy of wiring. It may cause failure of some electrical equipment or of the control and indicator circuit. Even worse, a chafed and often bare wire can ground out or short to another wire and start a fire. You cannot run to escape fire in a boat—you either must stay and fight it or go over the side. It is far better to keep it from happening by at least preventing chafe of electrical wires.

All wire must be secured into position using one or more of the various types of wire clamps available. Nylon clamps are among the best because they are tough, naturally insulated, and virtually impervious to most of the problems to be found aboard boats. These clamps come in all sizes, and should be placed about every 18 inches along a given run of wire or cable. Wire and cables affixed with clamps will not chafe other wire; and should there be any inadvertent strain on the wire, the clamps will help to prevent damage.

When a large number of wires are run down the size of the boat, generally hidden under the gunnels, they should be bundled and held together with nylon cable straps. These straps are inexpensive and very easy to use. The pointed end of the strap is inserted into the eye and pulled up tight using a pair of pliers.

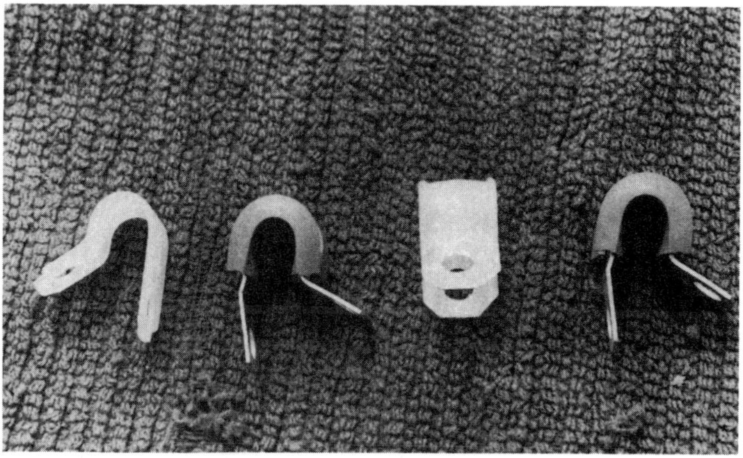

Fig. 1.10 White nylon cable clamps are best. Rubber-covered metal can be used but will eventually rust and cause trouble.

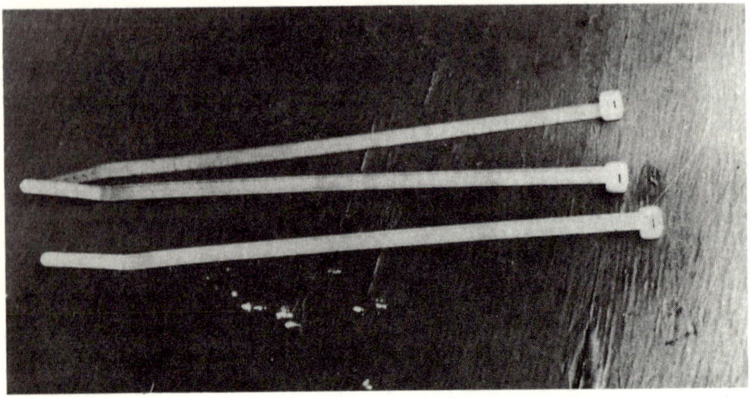

Fig. 1.11 Self-locking nylon cable straps will hold bundles of wire in place.

Fig. 1.12 Rubber feed-through grommets help to prevent chafe and short circuits in wire that must be passed through bulkheads.

The eye has locking teeth and will hold the strap tightly in position; and the excess tail of strap can be cut with a knife or diagonal cutting pliers.

From time to time it is necessary to pass a wire or wires through a partition or bulkhead. If the partition is of a hard material such as metal or fiberglass, the feed-through hole must be lined with a soft rubber grommet to prevent chafing of the wire when the boat hull flexes as it is underway. There are a variety of grommets in all sizes available in electrical supply stores.

On larger boats it is often necessary to run wire and cable through areas such as bilges that are more than normally hostile

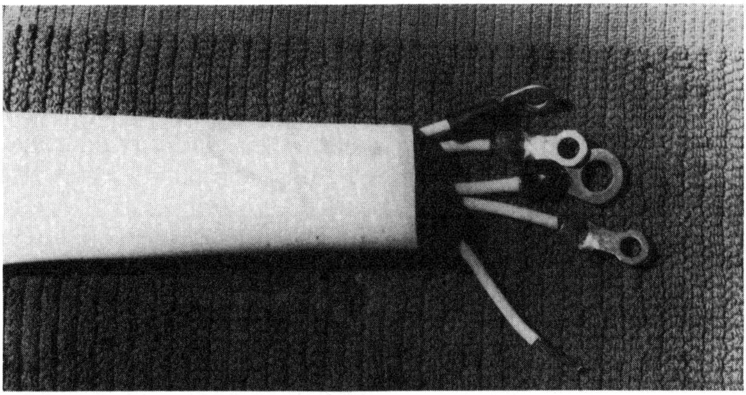

Fig. 1.13 PVC piping and flexible tubing both make excellent conduit for running multiple strands of wiring on small craft.

to electrical wiring. When this is the case you can use polyvinyl chloride (PVC) piping as a conduit. The cable or wires are first made up and passed into and through the plastic conduit. Any needed fittings are then glued into place on the conduit and the whole assembly is fixed in place with large nylon clamps. Lengths of plastic garden hose can be used in the same manner but should always be clamped into place. Wire in conduit is protected from all but the most severe trauma.

Wiring Diagrams

From time to time it may be necessary to refer to the wiring diagram for one or more of the electrical systems on your boat. When troubleshooting a faulty system or installing a new piece of equipment, the wiring diagram can be of substantial use. However, diagrams for some of the more complex systems can be somewhat difficult to read and interpret because of the number of symbols used and often the complexity of the system itself. (See table of diagrams and meanings in Chapter 6.)

A good wiring diagram will provide vital information such as the color code and size of the wire used in an AWG number. Values of current and voltage and fuse ratings may be given. Polarity directions, which are most important, along with tie points (where wires join) and ground or negative return points will appear on the diagram.

You must not expect the manufacturer of a production boat to provide wiring diagrams in the standard and more routine owner's manual. It is fortunate that, in general, the manufacturers are happy to sell you a copy of a more advanced maintenance manual from which you can glean nuggets of valuable information. Normally, the wiring diagram for the engine can be found in this manual. Very rarely, however, will you find a wiring diagram for the more routine lighting and other service circuits. If you have the time and patience to trace out these circuits, you will find that drawing up an equivalent wiring diagram will pay off the next time you are running down a shorted or open circuit. Be sure to include all of the information such as wire size and color, junction and hook-up points, fuse locations and ratings, voltage levels, and polarity indicators. Any data that will help to quickly trace and locate a wiring fault should be included. Don't try to make a pretty picture at first—just collect the information and scratch it down in rough draft form. Later, with the proper drawing tools and maybe some graph paper you can make a neat and informative drawing.

Fuses, Fuse Blocks, Circuit Breakers, and Switches

Fuses

Every electrical circuit aboard a small craft must be protected from overloads and shorts. Keep in mind that the primary purpose, especially on boats, is the prevention of fires and explosions. The secondary purpose is the protection of the equipment and the circuits themselves.

Normally, this protection is provided by fuses. On small craft the greatest majority of fuses are the glass Buss type, which come in a variety of sizes and electrical ratings (see examples). On larger boats you will find the more elaborate systems of circuit breakers and electrical control and monitoring panels; on small craft simple fuse blocks (which in themselves are distribution points) equipped with small glass Buss fuses, do very nicely.

It will be helpful if you understand how a fuse does its job and some of the factors that can prevent it from protecting its assigned circuit.

A typical small glass Buss fuse is a special metal strip placed within a glass tube for safety. The ends of the glass tube are capped with metal contacts and the fuse material is electrically bonded to the caps. All metal has a certain amount of electrical resistance, and current passing through this resistance in the metal strip can be thought of as energy. Some of this en-

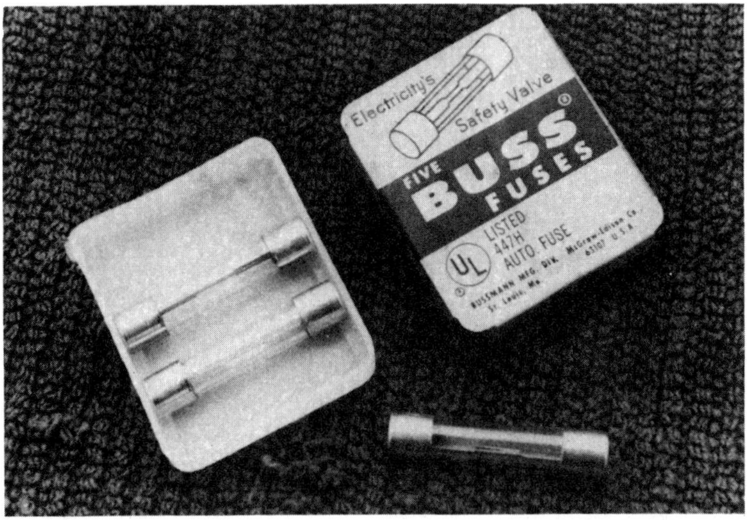

Fig. 1.14 Small glass Buss fuses are the main circuit protection on most small craft.

ergy is converted to heat. Excess current through the fuse metal link means excess energy and excessive heat. Enough heat will cause the special metal of the fuse to melt. However, *melts* implies a somewhat slow action, but in fact at some critical temperature the metal of the fuse link flashes into a gas—a far more rapid action.

The Buss fuses used in small craft electrical circuits are little different from those used in automotive circuits. As a result, they are vulnerable to the action of moisture, air, and salt. The metal used in the end caps of the fuse is, and has to be, a good electrical conductor. As such, it is an active metal in a chemical sense and will readily allow oxygen atoms in air and water to form undesirable oxides on its surface, which can, and often do, cause severe electrical problems.

These fuses come in several sizes, and although all are similar in shape they can be very different in terms of their electrical rating. There is a different current rating for each application. Some fuses have a tiny spring inside of the glass tube that applies tension to the fuse link and accelerates the disconnecting action of the fuse as it "blows." Other special types are designated as "slo-blow" and are used to protect a circuit in which

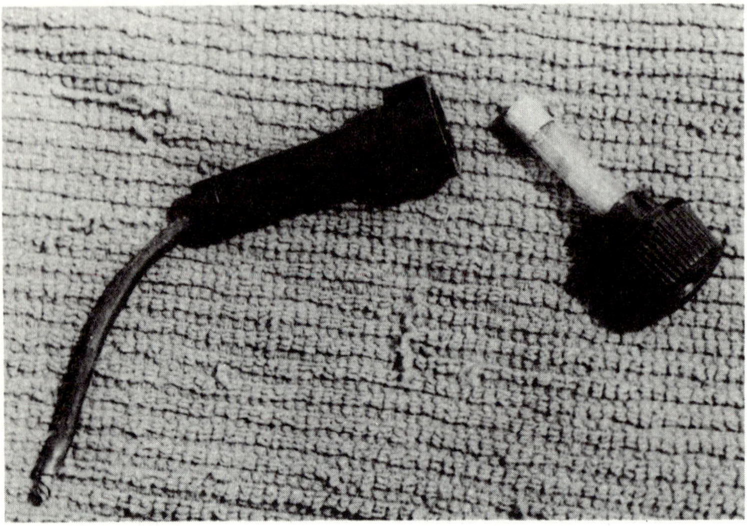

Fig. 1.15 Single circuit fuse holder or main line fuse holder (wire is missing on cap end).

the load is a motor of some sort and the initial demand for current is high.

Fuse holders for glass buss fuses also appear in a bewildering variety of styles, with the most common type being the black single fuse plastic holder used to protect only one circuit. Normally, it is installed directly in the line feeding the circuit it is to protect. Equally as common (at least for small boats) is the buss fuse circuit board of one or more circuit capacity. The use of these circuit boards allows the small boat skipper to add a number of fused circuits to his boat. The circuit board is not difficult to install and wire if a few simple and safe steps are followed.

Fuse ratings Care must be taken when selecting a fuse to protect a new circuit or equipment against overloads. A small glass Buss fuse can be found for almost every application on small craft, so you are not limited. Nearly all circuits will be of the 12 volt DC type so there will not be a problem in selecting the proper voltage rating. The physical dimensions will be dictated by the size of the fuse holder that is already installed or that you are getting ready to install. Most important is the blow-out rating, which you must observe with care, for even though the fuse has a glass tube and you can view the fuse link clearly it will tell you little. For the most part, the rating is etched in the metal end cap and is not always easy to see. Spare fuses should be kept in their original containers with the rating clearly marked so that you can find an exact rating duplicate of a blown fuse when you need it quickly.

To illustrate the importance of proper rating, let's assume you are installing a new radiotelephone transceiver. During the time the equipment is on and the receiver is active, the equipment draws 1 ampere of current at 12 volts DC. However, it would not do to use a 1 amp fuse to protect the equipment, for when you press the transmit button the current demand shoots up to 5 amps. Thus the fuse is selected for the maximum and normal amount of current that the circuit is expected to consume in regular operation. A second and equally important consideration for fuse selection is the size of wire in the circuit that the fuse is protecting. In no case should the fuse be current rated greater than the current rating capacity of the feed wire. Use the wire gauge and current capacity table to make sure that the fuse and wire are matched.

You will note that it is customary in small craft to fuse only one side of the power line to any circuit, normally in the positive side of the line. Often the manufacturer has installed an internal fuse of his own within the chassis of certain equipment, especially equipment that is polarity sensitive, as are nearly all types of electronic equipment. A circuit is so arranged within the equipment that if the power leads are inadvertently reversed the internal fuse will blow and thus protect the solid state devices from irreparable damage. When the internal fuse of such equipment blows it is reasonable to assume there has been an incorrect polarity of voltage applied, and the circuit should be carefully checked in a manner we shall describe later in this chapter.

Never replace a fuse with one of greater current rating than the circuit calls for. Remove the cause of the excess current demand if the proper size fuse blows a second time. Using an excessively rated fuse might cause a fire and may damage the equipment in the circuit it is meant to protect.

The advantages of glass fuses are: low cost, easy installation and replacement, ease of storage for spares, and reliability in protecting the circuit.

The disadvantages of glass fuses are found in the end caps and the spring loaded fuse clips that hold them. As previously pointed out these metals are chemically active and can easily corrode in a wet and salt atmosphere, resulting in a mysterious and gradual loss of voltage over a period of time. Further, the corrosion is not readily evident from a casual inspection of the fuse unless it is fully removed from the fuse clips so that the entire area of the metal contacts can be seen and burnished clean as necessary.

Circuit Breakers

Circuit breakers available for use on small craft operate on either a magnetic or thermal overload principle. Of the two, the magnetic seems to be best as the thermal types can be affected by external temperatures that disturb their set points. For example, on a hot summer day the temperature in the cabin (where the breaker panel is located) can be high enough to cause a thermal-type breaker to operate prematurely. The magnetic types will "trip" out (open) only when the load current is beyond their rated value. Many electronic technicians feel that the action of circuit breakers is too slow to fully protect modern marine electronic equipment. Other less demanding electrical gear will be fully and adequately protected by circuit breakers.

Perhaps the greatest value of circuit breakers as opposed to fuses is that they can be easily reset to the on position once they have been tripped by a momentary circuit overload for example, you might be considering installing a 12 volt windlass to haul the anchor. A unit of this type will draw electric current proportional to the load on the motor, which can be very heavy from time to time. You would find that you had to replace a fuse, even one of somewhat heavy rating, each time there was an extra heavy anchor-hauling event. In this case a resettable circuit breaker would be much more practical.

It you are considering the installation of one or more circuit breakers aboard your boat, use only those that are designated as "trip free." With this type of circuit breaker it is not possible to hold the breaker in the on position under overload, which is the same as placing an oversize fuse in the fuse holder. Trip free circuit breakers will continue to keep the circuit open as long as there is an excess current demand in the circuit being supplied, and the fault must be cleared before the breaker will stay in the on position.

Fig. 1.16 Example of trip free circuit breaker panel as used on Islander yachts.

Using a Test Lamp

Fuse panels and distribution boards on small craft are rarely located where they are easily accessible and where there is light to work by in order that a blown fuse can be readily identified and replaced or a breaker reset. Removal of each and every fuse for close examination in good light is not always an easy task, for the spring clips are designed to hold the fuse in tight contact and prying the fuse from its clips can shatter the tube. It is far better to use a test lamp to quickly check every fuse in the fuse block. Troubleshooting for a blown fuse can be further accelerated if each fuse is labeled.

Using a 12 volt *low* current test lamp to locate a blown fuse is not difficult. Just momentarily touch each end of the fuse clips with the leads of the lamp. If the fuse is OK, the lamp will not glow. When the lamp completes the circuit across a blown and therefore open circuit, the lamp will glow and indicate that either the fuse is blown or the end caps have become so corroded that they are, in effect, an open or high resistance circuit.

Keep in mind that the fuse link is shaped so that the blow point is in the middle of the link so that it will show as melted and therefore open; however, it will on rare occasions separate at a point that is not so readily evident. If you suspect that this is the case and the test lamp still is indicating an open circuit, it is best to replace the fuse.

ELECTRICAL SAFETY ON SMALL CRAFT

If we small craft skippers have anything going for us it's the fact that we are exposed to little personal shock hazard. Naturally, this statement is restricted to small craft in which the only electrical supply is the 12 volt storage battery and dock power has yet to come aboard!

It is most important to keep in mind that it is electric current that burns and kills. A bit of wry humor associated with this fact points out that there are old electricians and there are bold electricians—but there ain't any old, bold electricians!

Normally, the human skin resistance (25,000 ohms) is sufficient to protect you from both shock and current when you come in contact with no more than 12 volts of DC, assuming that the skin is *dry*! However, this is not to say that there is absolutely no danger from shock anywhere on the boat, for there is! Any contact with a spark plug wire or the high tension wire from the coil can give you a 14,000 volt "zing." Yet the voltage, as high as it is, in not lethal; but it is dangerous in another way. A brief contact with this voltage naturally will make you jump and pull your hand away and possibly cause a severe cut or bruise. Worse, if you are holding a tool at the time you may drop it into the engine, where it might short out on a battery terminal, cause an arc, and start a fire or explosion. Although the voltage is very high, it comes in pulses that are not long enough to cause injury. Moreover, the current is sharply limited. Nonetheless a shock from the ignition system can ruin your day! If you must service the ignition system while it is hot, be sure to read and follow the suggested techniques in a later chapter.

The internal voltages of some electronic equipment may be raised by special circuits designed for that purpose—yet another reason not to trifle inside the case of that gear. Normally all the lighting and other circuits are the same 12 volts throughout the boat and can be safely worked on if you take elementary precautions.

The primary danger on small craft is not shock (as long as the only voltage present is 12 volts DC) but fire and explosion resulting from sparks and arcs of short circuits, which will ignite any flammable material—especially if the material is gasoline vapor or gas from leaks around the stove. If you must work on your electric system *hot*—that is, with the battery connected—keep this danger in mind and take the most careful pains not to inadvertently cause short circuits with metal tools. Whenever you can, remove the battery cables from the battery and tuck them aside before starting to work on the electrical system. In that way it is not possible to cause an arc by dropping or touching a hot terminal with tools.

There is another hazard that is causing much trouble these days. When the battery is on charge or is being "jumped" from a second battery to start the engine, it will vent violently explosive hydrogen gas. This gas is normally carried away, but if it is present and there is even the slightest spark at the terminals, the battery itself may detonate and spray everything with hydrochloric acid.

As long as the voltage on small craft is restricted to 12 volts of DC from the lead acid battery, danger of electrical shock is very low. As can be seen, most of the electrical hazard comes from fire and/or explosions started from electrical arcs and sparks and gasoline or other fuel vapors. It can, therefore, be understood why many of the devices on inboard gasoline engines are designed for the marine environment and that the substitution of automotive equipment on the engine is foolish and dangerous. Three such devices are especially important in that respect: the inboard marine distributor, the marine service alternator/generator, and the engine electrical starting motor. Each of these devices in normal operation have the potential of striking a spark that, if not further protected, could cause the ignition of explosive gasoline vapors in the engine compartment. All these devices have been made vapor tight so that they can do their job without danger to the skipper, his crew, and his guests. Should you find it necessary to remove and replace one of these units, do *not* allow anyone to talk you into substituting an automotive-type device, no matter how much you think you are saving. If you do, you will be inviting disaster aboard as one of your guests!

In summary, shock hazard to people exists only when standard house currents and voltages are brought aboard small craft from the dock or by way of an auxiliary generator set. Dock power should not be installed without the protection of an isolation transformer between the dock power and the boat's circuits. In addition, protective devices such as ground fault interrupters must be included in the design of any dock power that is brought aboard.

Lightning Protection

Boats *do* get struck by lightning; and while the Coast Guard does not have specific statistics, you read about such events from time to time in local papers. Lightning strikes aboard small craft can cause severe shock and bad burns. Protection is neither complex nor especially expensive. Masts or radio antennas located so that they are the highest point on the boat form a cone of protection for the boat provided there is a good conductive path down through the hull to the water. Normally, a grounding plate is installed externally on the hull and the grounding plate is connected with a heavy cable to the mast or antenna. Some radio antennas come equipped with what is called a "horn gap insulator" that allows the energy of the lightning bolt to jump across the gap and flow to ground through the cable and ground plate. A special grounding plate called Dynaplate is very easily bolted to the hull under water and does not cause excessive drag. This plate, when used in conjunction with a central grounding strap or buss, will further protect you from equipment chassis shocks, help to keep electronic noise down, and help to prevent galvanic action from stray currents should you install dock power.

TOOLS AND TEST EQUIPMENT

Fortunately for most small craft skippers only the tools normally carried in the boat box, plus a few specialized tools, are all that are needed for a solid planned preventive maintenance program. You should have one of the new combination terminal lug crimping tools that cuts and strips wire, which can be found in any good hardware, electrical, or electronic supply store. You should have an assortment of crimp-type lugs. 35 mm film cans will keep them sorted by size and application and protect them from exposure to air until you are ready to use them. The solderless crimp-type lug should not be used on small craft without adding a touch of solder, but you can always carry a small gas torch or delay the soldering part of the job until you get back to the dock where you can use the electric soldering iron.

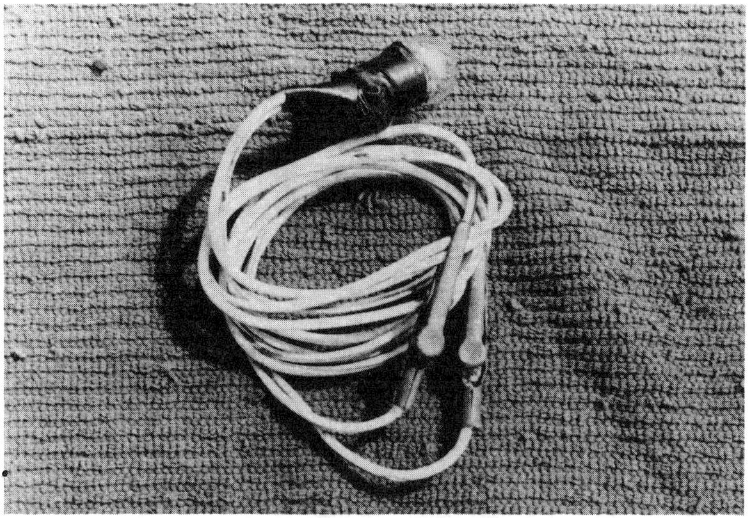

Fig. 1.17 This test lamp is not very pretty, but it does the job and the price is right!

Even though the crimping tool often comes with a set of jaws for cutting wire (in addition to all the other things it will do), you will find a good rugged pair of diagonal cutting pliers handy. You may have a pair aboard already, for they find much use in the removal and installation of the numerous cotter pins used on small boats.

For troubleshooting 12 volt DC small craft problems you should make up and learn to use a 12 volt test lamp. Although the size of the lamp is not too critical, generally the smaller the lamp the better. The smaller lamp will light easily on an open circuit when the normal current load for that circuit is very small. A no. 1847 lamp should do nicely. As an added feature you can make your test lamp polarity sensitive by adding a semi-conductor diode in series with one of the test leads. Almost any diode will do that can stand the 12 volts of DC and the current level of the lamp. With the polarity-detecting test lamp you can determine the correct polarity of any pair of wires with twelve volts of DC on them and no proper color code. The lamp will light only when the correct polarity is applied to the test leads.

Fig. 1.18 This inexpensive test meter will measure AC or DC volts, ohms of resistance, and low levels of current.

Even more versatile than the test lamp is a small pocket-type multimeter, which can be purchased in most electronic hobby shops for under $20. This meter will measure the value of an applied voltage, will determine open and short circuits, and best of all will measure the resistance of a given circuit so you can run down a pesky corrosion problem in an electrical connection. It will also measure a small amount of DC current, but the upper limit values are quite small and often do not apply to small boats. Moreover the meter must be inserted into the circuit in such a way that the current flows through the meter, and such a hook-up is rarely ever convenient.

2
SMALL CRAFT BATTERIES

It may seem unusual to you that we are devoting an entire chapter to anything as routine and mundane as the lead-acid battery, but remember that the greatest number of calls for Coast Guard assistance were from boats that had an actual or apparent engine starting battery failure. There are a number of factors that provide us with the roots of this problem, and we shall devote this chapter to the practical elimination of them by the PPM skipper.

The first and most easily corrected cause of potential or actual battery failure can be found in the design of the majority of small craft batteries. Very few stock boats, no matter what the propulsion persuasion—inboard, outboard or inboard/outboard—are delivered equipped with a *marine service* lead-acid storage battery. Most have a standard automotive surface charge battery. When in a motor vehicle, the battery is called upon to do but one job—start the engine. It has to provide only a brief but large surge of current to the starting motor. The subsequent electrical load is assumed by the highly efficient alternator.

This process, of course, also takes place in small craft, but only as long as the engine is running. And, as any cruising skipper knows, there are many times that the engine is not running but several electrical loads are still drawing energy from the battery. The automotive-type battery simply does not have the guts for that kind of service.

Today's 200 horsepower outboard engines cannot be started with an emergency starter cord as the 75 horsepower engines could in years gone by. Even when the battery was too low

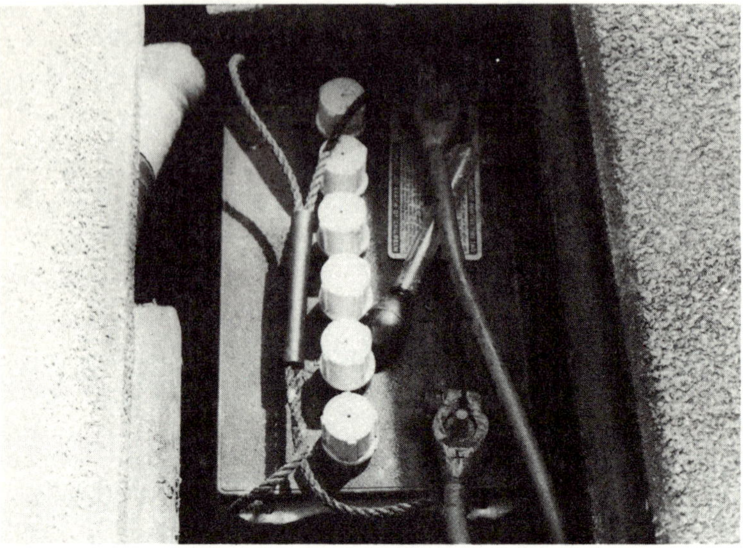

Fig. 2.1 The small craft marine battery—below decks and out of sight, out of mind. Not very glamorous or dramatic but essential!

to kick it over there was still a little juice left in it to provide a spark if you could twist its tail with a starting rope! Of course there is no other way to start almost any inboard engine except with a fresh "hot" battery.

So much for battery design problems. Now for the real reason that batteries in small craft fail so often—lack of the most primitive preventive maintenance! We have a couple of theories of why this is so. In almost all cases the battery is under decks—in a compartment and out of sight. It follows, does it not, that out of sight, out of mind. For this reason the battery will rarely get any kind of maintenance until it needs *corrective* maintenance, which, because of its construction, is almost impossible. You cannot replace broken or worn out parts to put it back in operating condition, but must replace the whole battery.

Another reason for the lack of battery maintenance may be that the battery lacks drama. That is, when it's doing its job, there is no noise, no light flash, no wheels going around—nothing! It's just a big black lump. Thus the battery tends to be ignored until it is dead or near dead. Things that have more drama when they are running tend to get much more attention from maintainers.

Also, there is considerable confusion as to just what a battery is in the minds of many small craft skippers. There may be an aura of mystery about what it does and how it does it. Many skippers still think that the lead-acid storage battery stores electric power. It does not! It stores chemical energy and by a fairly simple process, a phenomenon known about almost from Ben Franklin's day, converts the stored chemical energy to useful electric energy *only* when an external load is connected to it.

It is often convenient to think of a storage battery in terms of a watch spring. Wind up the spring and the energy stored in the spring will cause the rest of the watch movement to function. When the spring has wound down and expended its energy, it must be rewound or nothing happens. This analogy is applicable with one exception. The battery, unlike the wound-up spring, is a chemical device—*actively* chemical. The active materials, the lead plates and the acid, tend to keep on working even when there is no load on the battery. As a result, a fully charged but unused battery that sits unattended will eventually lose the stored chemical energy or *charge*. On small craft this gradual leaking away of the stored charge is very often accelerated, even when the battery is fully charged from the last trip, from various leakage paths provided by dirt and corrosion on the top of the battery and by external circuits that have formed leakage paths from corroded wires and switches. The wise PPM skipper finds a way to have the engine started and warmed up for a few minutes each week even when the boat is not actively in use and thus keeps the battery ever ready!

SPECIAL DESIGN FEATURES

A well designed marine service battery will have a more rugged case than one built for automotive service. It must be able to stand up to the pounding, vibration, and hostile atmosphere found in the bilge areas of most small craft.

A marine service battery will have a significant number of extra rugged plates over that of the automotive battery. Generally, the greater the number of plates and the larger the plate area of active material that is exposed to the action of the electrolyte (solution of acid and water), the greater the energy-delivering capacity of the battery. A number of misleading claims are made by advertising copy writers in their efforts to promote their batteries. *Every* 12 volt lead-acid battery will have exactly the same open circuit terminal voltage when fully charged. Ig-

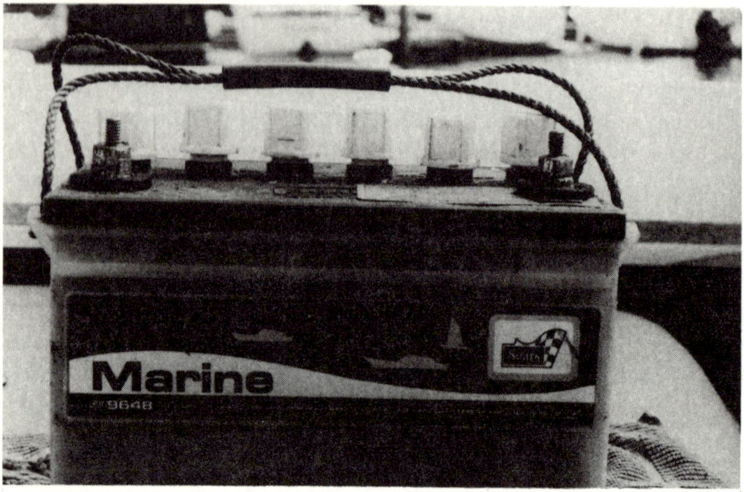

Fig. 2.2 Example of a small craft marine service battery. The Sears battery is modestly priced and has a number of useful features: carrying straps, threaded terminals for extra connections other than starter cables, and a rugged plastic case.

nore any claims that a given battery has higher voltage. Even the ampere-hour (AH) rating can be misunderstood. This rating is no more complex than the product of the discharge current times the number of hours the battery can sustain the rate. The discharge rate in amperes, the temperature of the battery during the discharge, the specific gravity of the electrolyte, and the final voltage at discharge all affect the ampere-hour rating.

AH ratings are based most commonly on a standard eight-hour period. Thus a 200AH battery could deliver a continuous load of 25 amperes for eight hours before its terminal voltage dropped to an unusable level. As a matter of fact, you will find that battery costs seem to rise in proportion to the AH rating. On small craft the greatest and heaviest load that the storage battery must provide is, of course, the huge but momentary demand of the electric starting motor. For a new, cold, and tight engine, this demand is often as high as 100 amperes. Usually the demand is limited to a very short period of time, and the alternator immediately takes over the balance of the electrical loads while at the same time restoring the energy drawn from the battery.

Some small craft skippers are surprised to discover that the efficiency of a marine battery located in the relatively hot engine spaces, where the temperature may be as high as 120 degrees, increases over 15 percent.

A battery filled with a greater concentration of sulphuric acid will have a higher specific gravity but it will also have a shorter useful life and will lose its charge faster when standing idle.

A marine service battery will be similar to the deep discharge type used in golf carts and electric fork-lift trucks.

A marine service battery will often have flat instead of round terminals. The flat terminal allows for a greater contact area to carry heavy currents and makes cleaning and burnishing much easier. Some of the better but lower cost units will include a screw-threaded terminal atop of the starter cable terminals to permit the attachment of other electrical loads directly to the battery—a most useful addition.

Built-in carrying straps and a handle are extra-sensible design features because the marine battery, unlike a car battery, is often removed and replaced in the boat. The marine battery, because of the extra weight of more lead plates within, cannot be safely carried using automotive-type terminal carrying straps. The battery may be so heavy that the terminals may pull out of the battery. If you must carry or move your marine battery, lift it by the bottom and observe the rules for lifting heavy objects to save muscle strain!

Many skippers will not find the newer automotive "lifetime" batteries suitable for small craft use. Although they require less maintenance, they are still shallow discharge batteries only suitable for starting loads.

INSIDE THE MARINE BATTERY

When a lead-acid battery is first formed at the factory both the positive and negative plates within each cell are composed of the same material—a soft spongy lead. These plates are subjected to a process that changes the positive plate to a brown metal called lead peroxide while the negative plate remains chemically lead. A lead-acid battery is composed of a number of cells, each of which has a terminal voltage of approximately 2.0. These cells are internally connected in series. A 12 volt marine battery will have six cells connected in series. Thus, every bit of electric current that flows from the battery to an external load must first pass through each and every one of the cells. If even one cell should fail, the entire battery is usually permanently disabled.

At Charge

A lead-acid battery, when charged and ready to go to work, will have a diluted solution of sulphuric acid and water, or electrolyte, completely surrounding and covering the plates. Because the mixture of acid and water is somewhat heavier than plain water, a specific gravity gauge is used to measure the state of a battery's charge. Normally, at full charge the specific gravity of the electrolyte is about 1.260 and will vary only if the temperature of the electrolyte changes. But more on that later.

So far only chemical energy has been stored in the battery. However, when an external electrical load is connected to the terminals, an electric current will flow from the battery. Modern electronic theory suggests that tiny electrically charged particles called electrons will leave the negative plate, flow through the external circuit, and return to the positive plate. It is of interest to note that for every single electron that leaves the negative plate, an equal number of electrons will return to the positive plate. Or, putting it another way, the current leaving and returning is precisely the same.

At Discharge

When the battery is fully charged the electrical pressure needed to pump the current is readily available. As the current flows the acid in the electrolyte gradually combines with the lead in the plates to provide the chemical energy that is converted by the action into electrical energy. Although it is a fine technical point, it can be said that the lead-acid battery does not store electrical energy—only chemical energy. As the acid leaves the electrolyte the terminal voltage or pressure gradually drops and the internal opposition to the flow of current (impedance) gradually increases until a point is reached at which the battery is said to be discharged. At this point nearly all of the acid has left the electrolyte and is combined into the lead plates. The specific gravity of a discharged battery will give a hydrometer reading of 1.135 at 77°F.

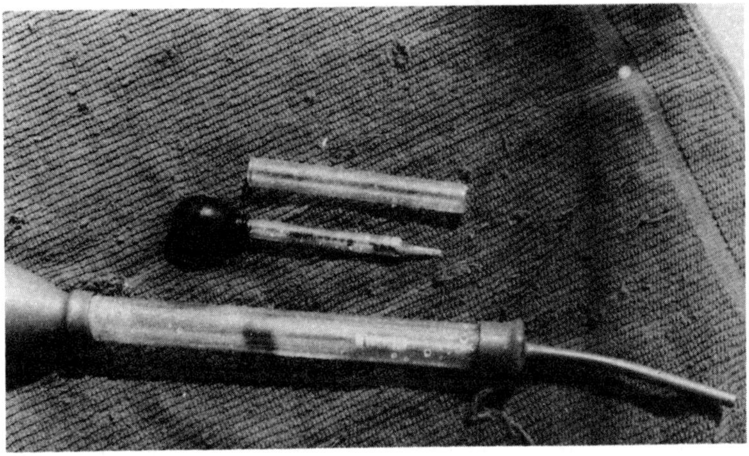

Fig. 2.3 Both of these hydrometers are inexpensive and will do the job. The floating ball type (top) can be stored on top of the battery when not in use.

Specific Gravity and the Effect of Temperature

A PPM program for small craft batteries must include the storage and use of a small and inexpensive hydrometer aboard. Learning to read and interpret the hydrometer is child's play. Since a lead-acid battery is a chemical engine, and since chemical activity is affected by temperature, the specific gravity reading that you will observe on your hydrometer may be in error unless you compensate for the temperature of the electrolyte. The center point or reference temperature is 77°F. For every three degrees of temperature above 77° add one gravity point. In the same manner subtract one gravity point for each three degrees below 77°F. For example, if you check a fully charged battery on a hot day in a hot engine compartment where the temperature of the electrolyte is as high as 107°, the hydrometer might read only 1.235, which may seem a little low. But when the correction factor for the higher temperature is added, you come out with a specific gravity reading of 1.260—right on! Specific gravity readings taken right after water has been added to the battery will be false. You should not make any attempt to read the specific gravity of a charging battery until it has been on charge for a while or the reading will again be in error.

Battery Terminal Voltages

There are two measures of battery voltage: the open circuit or *unloaded* voltage (when a fully charged battery is sitting on the ground with nothing connected to it) and the *load* voltage. Normally, an unloaded 12 volt marine battery will have a terminal voltage of 12.6. When an electrical load is connected, this terminal voltage will drop slightly at first due to the internal impedance of the battery. If the load is left connected, the voltage will continue to drop to a point where it is too low to pump any more current through the resistance of the load. This voltage, called the *final voltage*, may be as low as 10.5 for a 12 volt marine battery. In this condition it will not turn over the starter motor on your engine but may still have enough oomph to light a small lamp. Overdischarging can cause the formation of sulphates on the plates that will damage the battery. A battery left in a discharged condition can easily be ruined, which is why we urge you to keep your battery on a trickle charger during the off season when the battery is not in use. It is of great importance that you understand that the battery must not be always under a trickle charge, for battery engineers feel that this tends to damage the battery also. However, it is a simple matter to check the specific gravity once a month and connect the trickle charger only when the battery requires it, unless you have an automatic charger that shuts itself off when the battery is fully charged. Your boat battery, if kept at home in the off season, up on wood blocks, and fully charged at all times, can also be used as a booster battery for starting a cold automobile engine.

Distilled Water for Batteries

For many years there have been spirited discussions concerning the need to use distilled water to top off the electrolyte in a lead-acid battery. No sensible skipper will contend that distilled water isn't the best, yet many will point out that they have used common tap water for years and have had no trouble because of it. True, in many cases plain tap water will not excessively harm the battery. It can reduce its effective life *if*—and that's a big if—the local water has various natural chemicals in it. For example, in a number of communities the water is loaded with iron, which is definitely not good inside a lead-acid battery. Distilled water can be found in any drugstore for pennies and can be kept indefinitely. When, through the natural action of evaporation and the charging action of the alternator, the water level drops enough to expose the top of the plates, it is time to top off the battery with distilled water. One skipper we know has a high quality Surrette marine battery which he has carefully maintained for eleven years. The battery is now in the third boat and shows no signs of deterioration. The owner says he expects to leave it to his grandson when he passes on!

If you have any doubts, call your local water department. In most communities they are required by law to periodically test the water and must know its composition. If you have been using tap water all along and have had no problems with battery life, then by all means continue to use it, but the use of distilled water will extend the life of a marine battery.

PREVENTIVE MAINTENANCE FOR BATTERIES

Installation

If your boat battery is already installed, you might check its location for safety and convenience. The explosive gasses given off during charging must be quickly carried off and out of the boat, which means that the battery MUST be installed where there is a good stream of ventilating air. On outboard boats the battery is very often kept in a plastic battery box out in the open, or nearly so, and any gasses are carried off by natural air currents. With inboard engines, however, the battery is usually below decks, and care must be taken to choose a location where it will get plenty of air.

The battery must be located well up out of the bilge so that there is no danger of exposure to salt water. Salt water and sulphuric acid combine to generate deadly chlorine gas. Also, salt water, rain, or spray should be kept from terminals and vent caps.

The engine starting battery should be placed as close to the starting motor as is practical. The shorter the cables the better. When cables are kept short the voltage drops across the cables are kept to a minimum. Perhaps of equal importance, the battery should be easy to get at for inspection and service, such as cleaning terminals and topping off with water.

Even in a new and first class installation it should be recognized that a battery will have to take a physical beating. When a boat is running at high speeds, pounding over waves and making sharp turns, the battery should be cushioned and held in place with straps and clamps that are not affected by battery acid, sea water or dampness. If the boat is to be trailered this is yet of more importance. The best of installations include a heavy and ruggedly constructed battery tray. The tray, usually constructed of wood, is either lined with lead or fiberglassed and is bolted to a frame or stringer in the boat. Heavy plastic straps or wood clamps hold the battery tightly in place in the tray. If the boat is aluminum, steel, or other material that might be affected by the battery acid, the tray becomes a must.

Wiring

When you are installing a battery or thinking of replacing cables that have rusted or have frayed and dangerous insulation, keep in mind that these cables must often carry a starting current of over 100 amperes. A full sized inboard engine, for example, should have 4-gauge battery cables at the minimum. A lighter outboard engine can do nicely with 8-gauge cables. These cables should be heavy duty, multi-stranded, and preferably plastic insulated.

Under no circumstances should a battery ever be center tapped to pick off a lower voltage. Thus, a 12 volt, 6 cell battery should not be center tapped between the third and fourth cell to energize a separate 6 volt circuit. The 6 volt circuit should be supplied by a separate 6 volt battery.

Fig. 2.4 New cables must be made up with care. Use fresh cable and new connectors.

Terminals

The battery terminals and the engine "hot" terminal and ground terminals should be chosen with care. Pick the terminals with the greatest relative surface area for the size of your cables. Avoid screw-type clamps, which cause nothing but trouble for they rapidly corrode in a boat.

Strip the insulation from one end of the cable only enough to allow the bare wire to fit into the terminal socket. Twist the strands slightly to initially bond them together. Then, using a small gas torch, flow solder into and over the strands of the cable. Before you solder the cable end to the battery terminal, slip a section of shrink tubing over the cable and slide it up the cable out of the way. Now insert the pre-soldered end of the cable into the terminal socket and solder the two together, making sure the solder flows onto the cable and fills the terminal socket. When the terminal has cooled, slide the shrink tubing up the cable to cover the area where the cable joins the terminal socket. Add a little heat and the tubing will shrink tightly around the cable and the battery terminal clamp. This procedure will effectively keep air, salt, and other contaminants away from this critical joint. In the same manner, connect the stud end terminal to the engine.

Position the battery and install the cable for the shortest possible run to the engine, taking pains to clamp the cable every few inches with plastic cable clamps and making sure that there

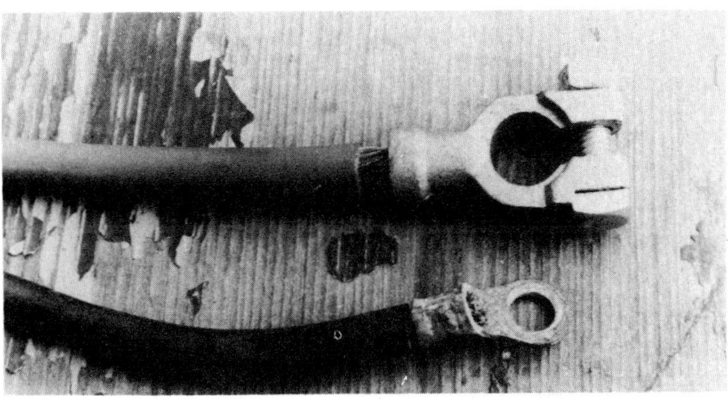

Fig. 2.5 Stripped of insulation, cable is "tinned" (pre-soldered) with rosin core solder, inserted into battery clamp, and soldered. Engine starter end is crimped and soldered also.

are not points at which the cable insulation could chafe through. If the cable must pass through any drilled holes to reach the engine, be sure to line these holes with rubber or plastic feed-through grommets.

Dual Battery Installations

In Chapter 6 we will describe the advantages and installation of dual battery installations. Naturally, extra cables are required, and you should follow the procedures just described if you are installing a second battery and battery transfer switch or electronic isolator.

Caps and Vents

Most marine batteries have battery filler caps that contain vents through which escape the gasses generated when the battery is on final charge. At the beginning of each boating season remove these caps from the battery, give them a thorough washing in a dilute solution of baking soda and plain water, and rinse. Check to make sure that the vents are not plugged, then dry and install them back on the battery.

Special vent caps containing a catalytic chemical that causes the gasses to recombine into water and drop back into the battery can be obtained from the Surrette Battery Company of Salem, Massachusetts. These vents are well worth the investment because they almost eliminate the need to add water to the battery during the season.

Fig. 2.6 These special Hydrocaps will increase the life of your battery by returning battery gasses and water back into the battery.

Routine Preventive Maintenance

At least once a month, and more often if the boat is used heavily during the boating season, set aside about an hour to do some simple but extremely important preventive maintenance. If it can be accomplished without too much difficulty, remove the battery from the boat and give it a good bath in soap and water to remove dirt and grease from top and sides that would otherwise tend to form leakage paths. At the very least periodically wash off the top of the battery with plain water.

It is very important to remove the cable clamps from the battery terminals and inspect the terminals for signs of a white or greyish powdery substance. Various sulphur oxides, if allowed to accumulate in any appreciable amount—especially on the interface where the clamp is to be in direct metal-to-metal contact with the battery terminal—can cause serious trouble. These oxides are electrically non-conductive and are the primary cause of *apparent* battery failure, so that when you attempt to start the engine you hear just a clicking sound from the starting relay but no rapid spinning of the starter motor.

Clean and burnish the battery terminals and the cable clamps with a stiff wire brush or sandpaper and reconnect and tighten the cables. If you haven't already done so, use the blade of a screwdriver to cut a plus sign on the positive cable terminal and on the top of the positive battery post to remind you of the correct polarity of the connections. Even a momentary crossing of the cables will blow out the diode rectifiers of an alternator, which is expensive to replace.

Reconnect the cable terminals and cleaned battery posts and give them a light coat of Vaseline or insulating spray such as WD40 to help keep air from the terminals and prevent oxidation from forming. Chemically soaked pads, found in many automotive parts stores, will also help to prevent terminal oxidation.

With a good hydrometer, carefully check the specific gravity of each cell. Be sure to add or subtract for the temperature of the electrolyte, which should be very close to the ambient air temperature around the battery. If any one cell reads lower than the rest you can expect trouble from that cell in the future, and now is the time to think of replacing the battery. A fully charged battery at 70° to 77° should read 1.280 on the hydrometer, and 1.250 is the lowest acceptable reading. A fully discharged cell will read 1.150, and anything below 1.225 is considered very poor. Again, if any cell is lower than the rest, that cell is more than likely shorted and will not accept and hold a charge.

With the battery back in its normal position, you might do some routine checking around the alternator/generator and the starter motor and starter solenoid. Look for loose or corroded terminals and any signs of chafe or fraying of the insulation on the interconnecting wiring. Such checking takes but a few moments and can save a lot of time and headaches later.

Keeping the water in the battery up to the proper level increases the operating efficiency and prolongs the life of the battery. Periodically inspect and top off the cells with distilled water (30 cents per half gallon).

OFF-SEASON LAY-UP

If your boat must be taken out of the water and stored for the winter, or if the boat must go into a prolonged period of inactivity for any reason, the battery should receive special storage care.

Remove and thoroughly clean and rinse it, taking care that the solution does not get into the filler caps. Dry the battery off and store it in a dry area up on blocks of wood. For some reason, leaving a battery sitting on a cold cement garage floor seems to kill it. Perhaps it is the opposing chemical natures of cement and acid in the battery. If the battery is kept charged it will not freeze and can be used for jump-starting the family car on extra cold mornings.

About every six weeks check the state of charge with a hydrometer and place the battery on a trickle charger if indicated. Do *not* leave the battery on a trickle charger lest you damage the battery. However, if you have an automatic type of trickle charger that shuts itself off when the terminal voltage is up to normal, the battery may be left on the charger indefinitely.

3
GALVANIC ACTION

GALVANIC ACTION

For many skippers galvanic action, or electrolytic corrosion as it is often erroneously called, is a mysterious disease that attacks many of the valuable and indispensible parts of a boat. The root cause of galvanic action is the same electrical phenomenon that causes the lead-acid battery to work, which may be the reason that the term electrolysis is so often used. Briefly, when two dissimilar metals are placed close to each other and separated, but connected electrically by an electrolyte, the two metals may, and often do, behave like the positive and negative plates of a battery. A voltage difference will exist between the two and an electric current will flow. The direction in which the current flows is important, for the current will flow from the metal that is said to be less "noble" to the metal that is on a higher order of nobility. There is a simple and classic demonstration of this effect. Connect two strips of metal, one of copper and one of aluminum, and hang them in either fresh or salt water for a period of time. When you remove them, you'll find that the aluminum strip badly corroded. The aluminum is less noble and more active than the copper. The two metal strips may be called electrodes. The water is an electrolyte. An electric current flows from the aluminum to the copper and corrodes the aluminum.

The least noble metals are at the same the most active in a chemical and electrical sense. Magnesium, aluminum, and zinc are examples. They can be termed anodic and are electrically positive in nature. Highly noble metals such as nickel, Inconel, and No. 18-8 stainless steel are said to be passive, resist corrosion and are electrically negative. Galvanic action will take place in an alloy of common brass, which is used for non-marine wood screws that are no more than a combination of zinc and copper. When these screws are placed in or near sea water galvanic action takes place between the zinc and the copper; and soon the zinc is eaten away and the remains of the screw are little more than soft spongy copper with no holding power at all.

Every metal fitting and fastener that goes aboard a boat should be silicon bronze, No. 304 stainless steel, or No. 316 stainless steel. Everything else, sooner or later, will be damaged from galvanic action and in many cases will have to be replaced.

Since galvanic action is electro-chemical in nature, deterioration of less noble metals on the boat will increase during the warmer boating season and be less apparent during the winter lay-up. In addition, the water in which the boat operates is also a factor, because salinity and pollution cause increased galvanic action. There will be little or no galvanic action between two metals that are exactly the same.

Galvanic action cannot be eliminated but it can be controlled. Take pains to fasten that hardware that is going to be underwater with screws, nuts, and bolts of the most noble metal your pocketbook will allow. Another very sensible rule is to join underwater parts with fasteners more noble than the parts. In this manner the fastenings will be protected. Parts that are generally of heavy construction can take some galvanic action without danger. For example, it is better to lose a tiny bit of metal from a shaft strut than to lose all four holding bolts that keep the strut mounted on the boat.

Yet another effective action is to paint underwater hardware with a non-metallic vinyl or other plastic paint, which has the effect of insulating the electric conducting surface and inhibiting galvanic current flow. No matter how tempting, do not paint underwater metal fittings with antifouling paint. The weed and animal killing chemical in antifouling paint is usually copper or tin, which are both wildly galvanically active and will soon chew holes in whatever they are painted on. Thus the antifouling paint will not adhere well to the base metal and will be too busy being galvanic to stop marine growths.

Still another effective fix is to bond all of the underwater metal fittings inside the boat with a heavy No. 8 gauge wire. Run the wire from the rudder quadrant or post over to the strut mounts (standard inboard engine); then run the same wire up to a grounding brush that rubs on the drive shaft. In this manner any current will follow the lesser resistance of the heavy copper wire and not flow from the below-water parts and through the water.

When aluminum boats are involved, galvanic protection assumes yet greater importance. The first book of this series, *Hull Care and Repair*, discussed in detail the need for protection of the hull itself from metal-based bottom antifouling paints. Care must be used in mounting any metal parts to the hull of an aluminum boat when the metal part must pass through the hull. Bronze sea cocks and depth sounder heads or any parts constructed of metal that are dissimilar to the aluminum must be installed so that they are not in electrical contact with and hence a cause of galvanic corrosion of the hull.

Very little of this discussion applies to a skipper who trailers his boat unless he leaves his boat in the water for long periods of time. But if the boat is moored or sits in a slip, protection against galvanic action is an absolute requirement.

"Sacrificial" Zincs

Almost all skippers are familiar with the use of sacrificial zincs, which is the most common form of cathodic protection. Rudders, shafts, and even outdrives can be protected by bonding very active zincs directly to the metal parts. As galvanic electric currents flow from the zincs to other metals, the zincs are gradually eaten away and must be periodically replaced. When a zinc lasts longer than a season, you should suspect that it is not doing its job and try to find out why. In all cases the sacrificial zincs must be mechanically and electrically connected to the metal parts they are to protect. It will do no good to fasten a zinc to the hull of a fiberglass boat and not connect it, such as with a wire inside the hull, to the metal outdrive, drive shaft, rudder, or strut.

You should be aware than zinc anodes vary in quality. Inexpensive recovered zinc anodes will not do a good job and you will soon be replacing some very expensive underwater hardware if you try to cut corners. The best zinc anodes conform to military specification MIL-A-18001, which defines an alloy that will waste away at a very uniform rate and will not form an interfering crust on the surface of the zinc while it is doing its job.

Fig. 3.1 Examples of various sacrificial zincs used on small craft: the ball-like shaft zinc, the disc-shaped rudder zinc, and the honeycomb outdrive zinc normally mounted on the hull to protect the outdrive unit.

Make sure that your zinc anode has a cast-in galvanized iron or steel insert where the bolt fastens the zinc to the metal. The area around an unlined mounting hole will be eaten away and cause the zinc to become loose and therefore useless. If you are installing zincs for the first time and are not sure how big the protective zinc anode should be, keep in mind that a square inch of *quality* zinc will protect 800 inches of painted steel. It is best to err on oversize as a general rule, since anodes are not very expensive.

Take a walk through any boat yard in the spring and chances are you'll see at least one skipper carefully painting the sacrificial zinc anodes with bottom paint. The paint will act as an insulator and will greatly reduce the effectiveness of the zincs. Do *not* paint your zincs; and replace them each time you haul out to paint the bottom. Make sure they are well fastened, the right size, and are bright and shiny.

(a)

(b)

Fig. 3.2 (a) This sacrificial zinc has been painted over the bottom paint and is now useless. (b) Also shown is what is left of a 24-inch propeller left unprotected by a painted-over zinc.

Dockside Protection

Boats that are idle at dockside, tied to a pier, or moored out in the stream for long periods of time are most susceptible to galvanic corrosion. To increase dockside protection, you can purchase an extra anode cast in the form of a fish and connect it by wire to the engine using a heavy battery-type clip, which can be quickly removed. During dockside time the zinc end of the wire (or "Guppy" as it is called) is dangled over the side while the clip is connected to the engine block. Any galvanic currents will flow from the Guppy to the engine, and the regular galvanics will be further protected. This trick can prevent a mid-season haul-out to replace shaft and rudder zincs.

Protective Devices Against Galvanic Action

An electronic device is on the market that will sense the presence of a corrosion problem. It consists of two metal plates that are mounted on the hull and connected to a control unit within the boat. When the plates detect an upcoming galvanic problem, they actually impress a small current in the direction opposite to the galvanic currents and cancel them out. By far this method is the most effective of all and, of course, is also the most expensive. However, such a system will easily pay for itself when you have controlled all galvanic action problems in your boat. Mercruiser and OMC now offer electronic units that are designed to protect outdrives and larger outboard motors.

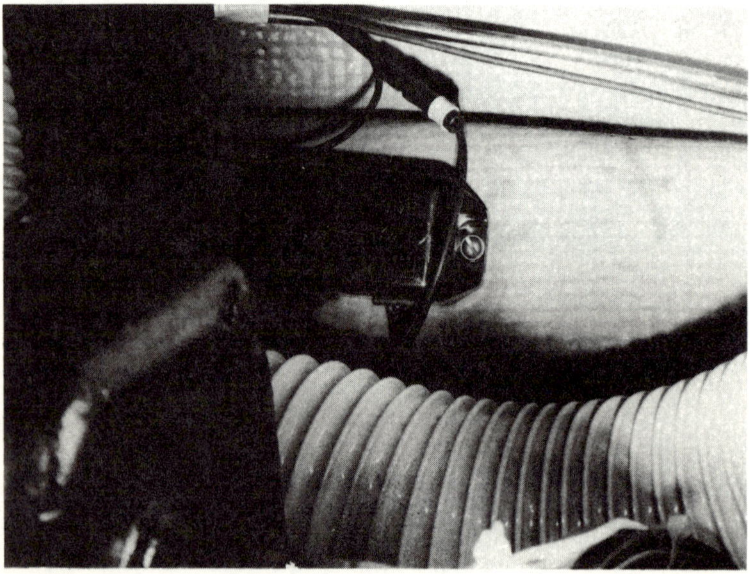

Fig. 3.3 What may be the ultimate in antigalvanic protection, the electronic Mercathode unit is mounted on the transom inside the boat. Not shown are the electrodes, which are underwater and on the outside of the hull.

An east coast mariner, at great expense, recently installed a retainer sea wall using heavy steel slabs pile-driven into the water's edge. A number of boats moored along the edge of this sea wall had severe galvanic action problems that they had never had before. In a few weeks sacrificial zinc anodes were eaten away by the stray currents flowing from the boats through the water and to the steel sea wall. One or two skippers installed electronic impressed current systems and solved their problems. The rest are replacing their zincs several times each season or have moved to better climes!

Stray Current Galvanic Action

Stray current galvanic action is by far potentially the most destructive of small craft and also the least understood. Most of the problems seem to stem from the boat's internal electrical system rather than from underwater galvanic action. As a case in point, the skipper of a large houseboat recently installed dock power. Within days holes appeared in the cases of two aluminum outdrives. Apparently, during the installation the ground connections were crossed between the boat and the dock AC power plug. A true expert marine electrician connected a voltmeter between the outdrives and the dock power grounds and discovered an AC voltage difference of nearly 5 volts! The galvanic currents from the boat through the salt water to the dock area were substantial and caused expensive damage. A simple rewiring of the dock power plug stopped most of the current but did not save the outdrives.

A 12 volt DC system will suffer stray current damage when the negative and positive battery leads are crossed between two electrical devices such as bilge pumps. A number of boats have a bilge pump forward and another aft. Water in the boat will connect these pumps and form a galvanic current path if the leads to one of the pumps are crossed. Although the pumps are not polarized (that is, they will operate even though the leads are crossed) a current will flow between them because there is a higher electrical potential in one pump than in the other. Always take the greatest pains to make all electrical connections the same polarity. That is one of the reasons for color coding the wires. Do not use scrap wire—get the correct size and color for all wiring jobs. You will save a lot of trouble in the future by keeping polarities straight!

SWITCHES, FUSES, AND CIRCUIT BREAKERS

Whenever you install a new piece of equipment that is to be powered from the boat's battery, make sure that the negative or ground wire is never switched or fused. Put all switches and fuses in the hot or positive side of the line. If the negative side is switched or fused and becomes open, a current may flow from the open connection through the dampness or water and severe galvanic corrosion may result.

4 GENERATORS, ALTERNATORS, AND VOLTAGE REGULATORS

It is most important that all small craft skippers understand that marine generators and alternators are unique. Unlike their automotive cousins they must be explosionproof and vaporproof. In a car there is little danger from the accumulation of explosive gasoline vapors for they either fall to the ground or are quickly blown away by the natural circulation of air. Not so on a boat, where drips or leaks fall to the bottom of the bilge and lay there just waiting for a spark. In normal operation both generators and alternators have points within them where there is present a normal amount of sparking. In the generator it is the commutator surface where the brushes make contact with the copper segments. In the alternator it is the brushes in contact with the field excitation slip rings. It is because of this sparkproducing potential that manufacturers are required to meet safety standards, and these standards increase the cost of marine-type machines. While it is true that automotive equipment can easily be substituted for a marine generator or alternator, the skipper who does so is trifling with his own and his passengers' lives.

GENERATORS

Since small craft still use the older type of DC generators, we will begin this chapter with some useful information concerning the operation and purpose of generators. It may be of interest to note that *all* generators or alternators start off by generating an alternating current! How this current is collected and treated depends on whether it is to be called an alternator or a generator. In any event the point is academic, for on boats, much like

Fig. 4.1 Example of older but rugged marine-type DC generator.

cars, both generators and alternators put out direct current. The output is first used to charge the battery, and second to provide electric power during those times the boat is underway.

The generator, often called a rotating machine, has three principal parts: a rotating part not surprisingly called the rotor, a stationary part called the stator, and a device for collecting the generator current and converting it from AC to DC called the commutator. Both the rotor and stator are large, rugged, heavy electromagnets. Heavy copper wire is wound in forms over soft iron cores. Very little trouble is ever experienced from these coils. Most of the problems in DC generators stem from the commutator area.

The commutator is a ring of copper segments insulated from each other with mica washers. Soft carbon brushes forced into contact with the commutator segments by pressure springs ride on the surface of the commutator both to switch and collect the generated current. It is here that most of the troubles with DC generators will be found. Brush springs break or hang up. Brushes wear out and the commutator itself gets coated with burned dirt, grease, and excessive carbon from the brushes.

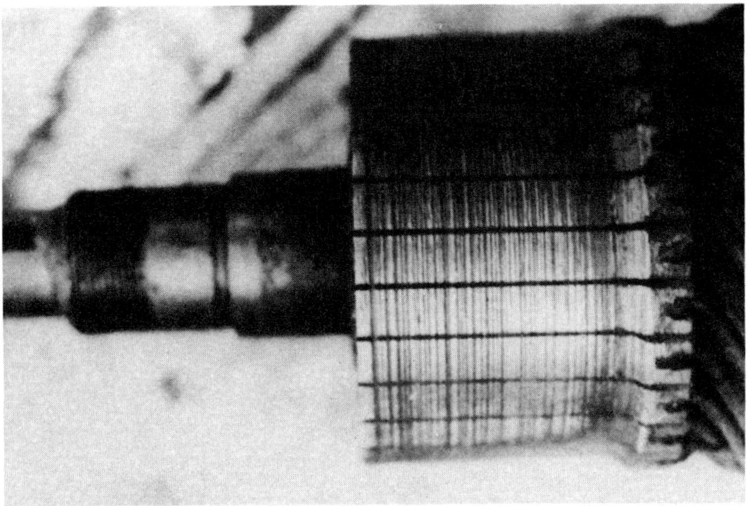

Fig. 4.2 The "soft spot" of the older DC generator is the commutator. Worn brushes have scored segments of the commutator; poor performance and much electronic noise in radios and depth sounders will result.

The rotor (also called the armature) in rare cases will develop an open or short circuit in one or more of the wire windings, which will cause it to lower the output current and to make despairing noises. There is no cure other than to remove the generator from the boat and send it to the repair shop.

Construction

The diagram shows a typical DC generator layout. Within the generator the two carbon brushes collect and commutate the alternating current to a direct current that charges the battery and carries the electrical load as long as the engine is running above a certain RPM. The major weakness of the DC generator is that at low RPMs it puts out very little current and will not provide much electrical power. Also, because of its design it is very heavy and large compared to the more modern alternator. Some skippers find annoying the excessive electronic noise caused by the action of the commutator, which must be treated to eliminate it from the primary circuits. Yet the DC generator is a rugged and reasonably fault-free machine and with the most rudimentary preventive maintenance will give reliable service.

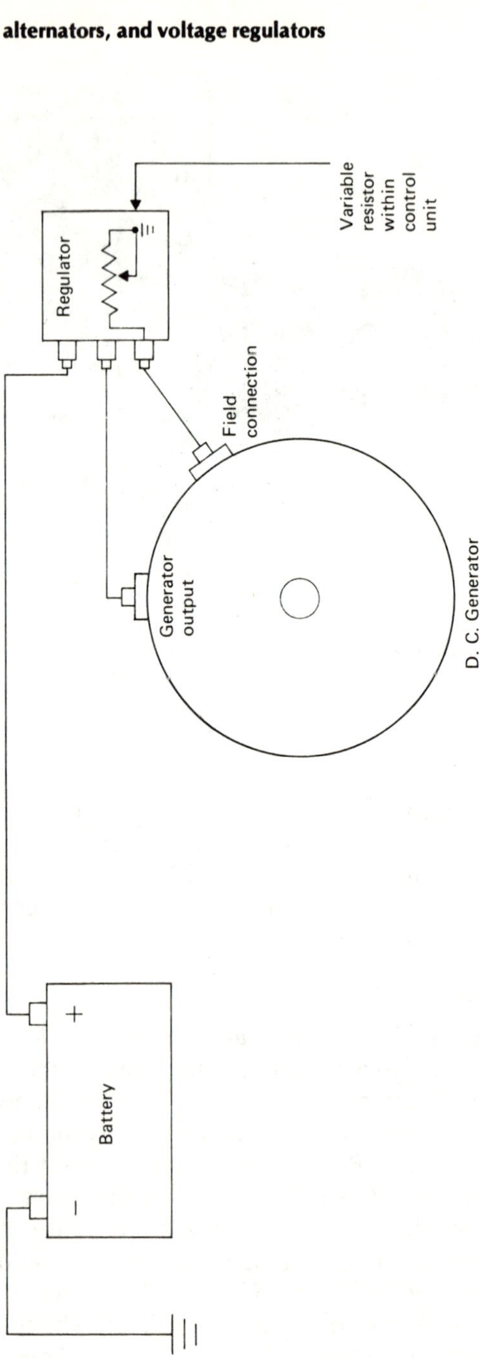

Fig. 4.3 Diagram of the DC generator showing various electrical connections.

Within the generator, one of the brushes is connected to the frame and therefore to the DC ground and is at the same electrical potential as the negative terminal of the battery. The second brush is connected to an insulated positive post, the output connection to the battery. A wire, usually of smaller size, connects the field excitation or control winding within the armature to the voltage regulator. Were it not for the voltage regulator the output voltage of the generator would vary wildly from zero volts at low or idle speeds to as high as thirty volts at high RPMs, which would burn out most 12 volt devices on the boat. The voltage regulator senses the output voltage and sends a signal current to the field winding of the armature. This current, called the excitation current, will cause the output current to the battery to increase or decrease according to a preset amount within the regulator.

In the DC generator there are other maintenance soft spots—the end bearings that hold the rotor armature shaft in place—but even these rarely cause trouble. On a few generators of the older variety you will find little oil cups in which a few drops of heavy duty oil should be inserted once or twice a season. The DC generator, like the alternator, is belt driven; and

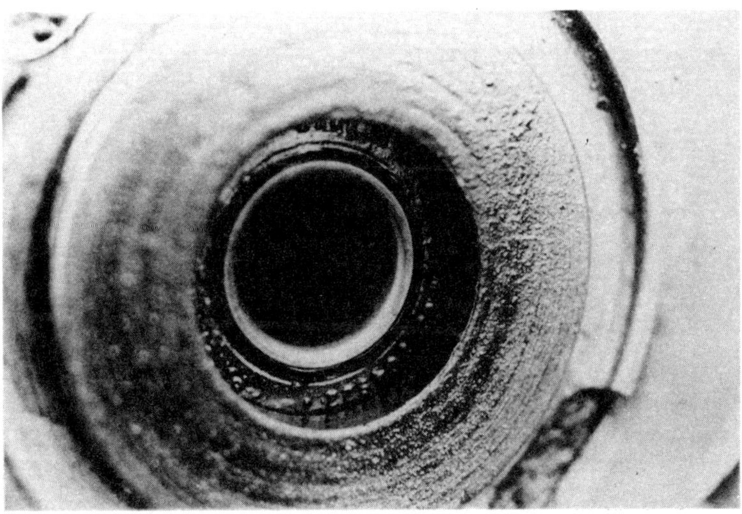

Fig. 4.4 This close-up view shows generator end bearings that must be renewed from time to time.

more problems stem from improper belt tension and wear and outright broken belts. Every skipper should carry at least one spare generator drive belt, if only to lend to a non-PPM skipper!

Voltage Regulators for Generators

The voltage regulator for the more conventional DC generator is an unimpressive black box, but there is a lot going on inside.

(a)

(b)

Fig. 4.5 (a) The inside view of a conventional voltage regulator shows three electromechanical relays: From left to right are the current limiting relay, the voltage regulating relay, and the cut-out relay. Note adjusting screw and contacts. (b) A back view of the relays shows the springs and adjusting screws.

Often there are as many as three electromechanical relays that control the output of the generator: the voltage control relay controls the generator output voltage; a current limiting relay controls the amount of output current; and a reverse current cut-out relay disconnects the generator from the battery when the engine stops. If the latter fails, the battery will almost immediately discharge itself through the generator brushes to ground when the engine is turned off. Each of these relays are spring loaded to hold them in a preset mechanical and electrical condition. The adjustment of these electromechanical positions can only be done by a trained and experienced mechanic using rather expensive tools and instruments. The average small craft skipper will wisely stay out of the voltage regulator except when conducting a simple test, to be described later, which will positively determine if the voltage regulator is a source of trouble.

The voltage regulator, like the commutator of the DC generator, is a natural electronic noisemaker. In normal operation the voltage and current-limiting relays switch on and off many times per minute, and the sparks appearing at the relay contacts make serious electronic noise in sensitive electronic equipment.

Generator Monitors

In some small craft an ammeter monitors the performance of the charging system. In most other boats there is an indicator light similar to those in automobiles. That neither the ammeter nor the light truly monitors the behavior of the system is not of great importance. When the indicator lamp is used a relay is provided to turn the lamp on and off, and this relay is a further source of noise and trouble. When the lamp is on the generator is not charging, and when the lamp is off it might indicate that the generator is charging or that the lamp is burned out! The ammeter is not cumulative—that is, it shows only the charging current to the battery and not the total electrical load current that various systems in the boat might be drawing. Few if any boats are delivered with a monitor voltmeter which will depict more accurately the action that is going on in the system. Skippers can purchase and install small voltmeters with little effort.

PPM FOR THE GENERATOR

Preventive maintenance for a conventional generator is extremely simple and requires little of the skipper's time. The

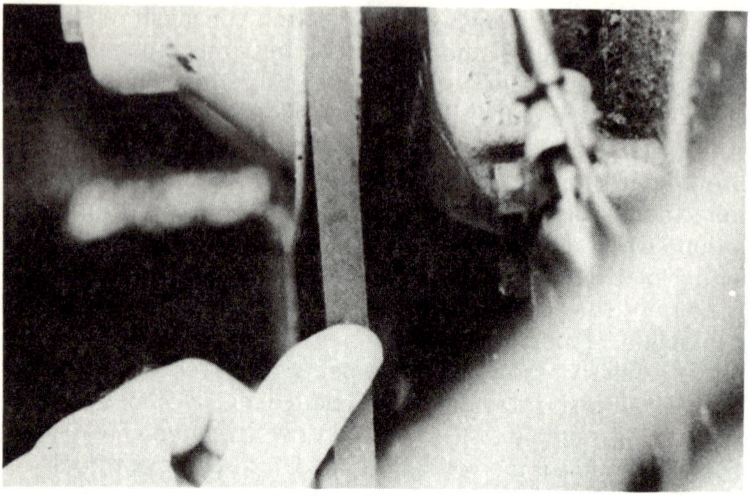

Fig. 4.6 Drive belt tension is correct when thumb can depress belt ½ inch.

first step is maintaining the drive belt in proper tension by adjusting the tensioning bracket when a finger test indicates the need. Drive belts stretch and wear and need attention from time to time. Overtensioning a drive belt is just as bad as undertensioning. A drive belt that is too tight will wear out and can cause excessive wear on the front bearing of the generator. To test drive belt tension simply press down with your thumb on the drive belt at a midpoint between the pulleys. A proper amount of tension will depress the belt about one-half inch. A slipping belt will make itself evident by a squealing noise and can be cured by the application of a small amount of belt dressing obtained at an automotive parts store.

In the early part of the season, before you launch the boat, take the drive belt off the engine and inspect it inside and out. Look for frays and cracks and a shiny worn surface. Replace a worn belt and keep a spare aboard at all times. If you have an outboard engine that uses generator drive belts, don't neglect them.

If you suspect that a generator bearing is about to let go, you can confirm by listening to the noise that the shaft in the bearing makes. Place a long screwdriver blade on the bearing housing and your ear on the handle of the screwdriver. The sound will be transmitted up the shaft of the screwdriver to your

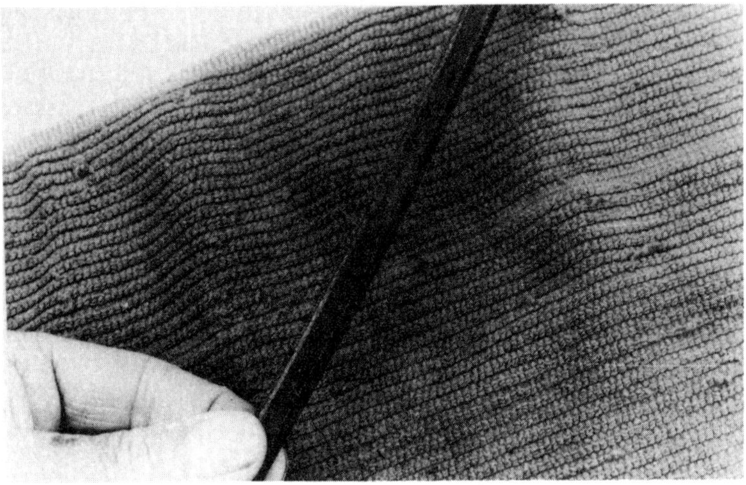

Fig. 4.7 Carefully inspect the inside of the drive belt.

ear. In this manner troubles from a bad bearing will be readily apparent. Most generators and alternators use sealed self-lubricated bearings and nothing short of replacement will cure a bearing problem.

Removal of a Generator

From time to time it may be necessary to remove the generator from the boat for corrective maintenance ashore. It's not a bad idea either to remove the generator for a long off-season storage. Dismounting a generator is not difficult and can be accomplished with a minimum of tools. There are, however, one or two precautions that you should take. It is always best to remove the battery cables before you begin. Tools do slip and will cause a very bad short circuit if they happen to fall against a positive hot terminal and any metal ground.

On the top of the generator there are up to three electrical wires: the positive wire that is connected to the battery, usually by way of the starter motor; the field winding connection that comes from the voltage regulator; and sometimes a third wire, which is a ground wire to the engine block. Each of these wires should be carefully removed and tagged with a bit of masking tape so that you'll know where they go when you reinstall the generator. The positive battery wire is of heavy gauge and usually is red or orange. The lighter-gauge field wire is usually green, and the medium-gauge ground wire should be black.

Loosen and completely remove the nut, bolt, and lock washer on the belt tensioning bracket. You may find it necessary to pry up the generator with a board or stick after you loosen the tension nut and bolt. Normally there will be one or two long bolts through the generator mounting bracket that can now be loosened and removed.

Removing an alternator is very similar but even greater precautions should be taken with the electrical connections. If you do not first remove the battery leads and inadvertantly allow the field winding lead to touch ground, you could ruin the voltage regulator, especially if it is a transistor-type regulator.

Generator Tests

If your ammeter reads discharge even with the engine running, the problem *could* be the generator but just as often is the voltage regulator. A simple test determines which it is. Start and run the engine at about 1,000 RPM. With a screwdriver or a jumper wire with spring clips, short (connect) the field winding connection on the generator to the frame or ground. Watch the ammeter. If the output soars to several amperes, the trouble is not the generator but the voltage regulator. *Caution:* Do not run the generator any longer than a few seconds. It is putting out maximum current and if it overheats may be destroyed.

Flashing the field

Older DC generators are able to start up on the small amount of residual magnetism that remains in the soft iron core upon which the copper wire is wound. Occasionally, however, and often after a long winter storage, the residual magnetism has left and the generator will not charge the battery with the engine running. All that is needed to get the generator back into good health is to give it a shot of good hot current directly from a well charged battery. Connect a heavy wire to the positive battery post and lightly brush the bare end of the wire across the output (positive) terminal. Do *not* hold the wire on the terminal—just brush it across the terminal once or twice. You should expect some sparks to fly, so make sure that there aren't any gasoline fumes close by. This brief but very high amount of positive current will remagnetize the field poles and the generator will now get back to work.

This same trick may be used when the generator has been wired into the boat with the field and output wires reversed, which polarizes the fields in the wrong direction and causes the generator to discharge rather than charge the battery. Simply connect the wires properly and then "flash the field" with the proper positive voltage.

Caution: Never try to run the engine with the battery disconnected or the wire from the generator to the battery disconnected, for this can easily damage both the generator and the voltage regulator.

If any of your tests indicate that the voltage regulator is at fault, there is very little that you can do short of removing and installing a new regulator. However, a trained mechanic can locate and eliminate some regulator troubles with special tools and instruments. Most modern voltage regulators have sealed cases that do not permit service even by a trained mechanic. All three relays, the voltage regulator, the current limiter, and the reverse current limiter are spring loaded and the tension can be adjusted for proper set points. As a rule, special voltmeters, ammeters, and proper bending tools are required plus the precise knowledge of how to use them.

Fig. 4.8 The sealed-case electronic voltage regulator cannot be adjusted or repaired, only replaced.

ALTERNATORS

Most new boats are equipped with a marine version of the automotive alternator that is vaporproof and explosion proof. The alternator does exactly the same job that the generator does except by design it does it slightly differently and substantially more efficiently. Most alternators are substantially smaller, use less heavy iron, and have less complex control circuits. They require very little service and, for the most part, will perform reliably with a small amount of PPM.

Construction

Essentially, the alternator on a boat, much like that on a car, is a three-phase AC generator—actually *three* generators in *one* case. Rather than having one power generating winding, as is the case with the conventional generator, it has three windings, which are wound and bolted into place on the outside of the rotor. This construction eliminates the need for heavy spark-making commutators and troublemaking brushes. The rotor is a simple electromagnet and contains the field windings that induce the voltage and current into the output windings. Two slip rings on the rotor feed the excitation current to the

Fig. 4.9 Example of an alternator rotor. Note field winding connected to slip ring on end of shaft. Fingers of iron create a rotating magnetic field. Hole is drilled by manufacturer to balance rotor on test.

rotor, and since this current is relatively small, there are few problems with the rings and their associated carbon brushes. The stator windings are connected to solid state rectifying diodes, which have the unique electrical property of passing current in only one direction. Thus, although part of the output of the windings is of the wrong polarity, the diodes do not conduct during that part of the cycle. AC within the alternator is converted to DC by the diodes and is then sent to the battery.

The three windings all working together enable the alternator to generate a substantial current even when the engine is idling—something the older DC generator will not do. Because the diodes pass current in only one direction, the need for any reverse current cut-out relay in the controller is eliminated. When the engine is shut down and the alternator stops rotating, the diodes effectively act as an open switch between the battery and the generator. In this manner the battery cannot discharge through the alternator. If there is a weakness in alternators it is with these rectifying diodes, which will burn out from time to time as evidenced by a lower output from the alternator often accompanied by a whining sound. The diodes are also very sensitive to incorrect polarities. If for any reason the wrong polarity should be connected to the output, the diodes will try to conduct heavy current in the wrong direction and burn out faster than a fuse. Use extreme caution whenever you are connecting up a new or rebuilt alternator to make sure that the proper wire is connected to the terminals.

"Dead Battery" Effect

A second but minor weakness of the alternator is that it will not work on a completely dead battery. The field excitation windings must have a small amount of working current if the alternator is to function. If you have a dead battery but can get your engine started, you can coax the alternator into operation by connecting a lantern battery or a half-dozen flashlight batteries in series to the field winding. The generated current will enable the battery to eventually take over the charge. The positive terminals of the batteries must be connected to the positive terminal of the field connection on the alternator.

Excitor current is prevented from flowing to the field winding when the engine is shut down by the connection of the winding lead to the ignition switch or by a simple relay that opens up the circuit.

Fig. 4.10 Close-up view of a vibrating relay in the voltage regulator. Contacts and reed can be seen just above adjust screw.

Voltage Regulators for Alternators

As we have seen, there is no need for a reverse current relay since the built-in rectifying diodes take over that function. Nor is there any need for a current limiting relay because the internal impedance (similar to resistance) limits the output current to a safe value. All that is needed is a voltage regulator. Electromechanical vibrating relay regulators are still being used, but they are gradually being replaced by the more modern transistor models. As is the case with DC generator voltage regulators, you can do little more than keep all connections clean and mechanically tight. When a voltage regulator fails it usually either runs wide open and overcharges the battery, causing excessive boiling of the battery electrolyte and severe gassing, or it won't allow the alternator to charge the battery at all.

Voltage Regulator Tests

To determine if the alternator voltage regulator is working, turn the engine ignition key on but do not start the engine. Go back to the engine and hold the metal shank of a screwdriver close to the back of the alternator. If the voltage regulator is working, you'll feel a magnetic attraction between the alternator field winding and the shank of the screwdriver.

Fig. 4.11 Illustration of a simple test of the voltage regulator for an alternator. Ignition key is turned on and screwdriver is held at back of alternator to test for magnetism.

A second but more positive test is the same as the test for voltage regulators on DC generators. With the engine running, connect a jumper wire from the positive output terminal to the field terminal and watch the reaction of the ammeter. The current charge rate should jump to a significant value. If it does, remove and replace the voltage regulator taking pains to connect the leads properly. In most cases solid state regulators are used with marine alternators. These units are sealed inside of metal cases and cannot be serviced—only replaced.

Trouble-shooting the Alternator Circuit

Far too often an alternator will balk because of a corroded connection or a broken wire. The alternator wiring on a boat can be somewhat complex, running as it does up to the operating position and back down to the engine. However, testing with a 12 volt trouble-shooting light is one of the easiest and quickest ways to find a bad or open connection. Once you have established that both the alternator and the voltage regulator are functioning properly, follow these steps until you locate the problem.

1. Connect one test lamp lead to ground on the engine and the second lead to the output terminal (orange or red wire). If the

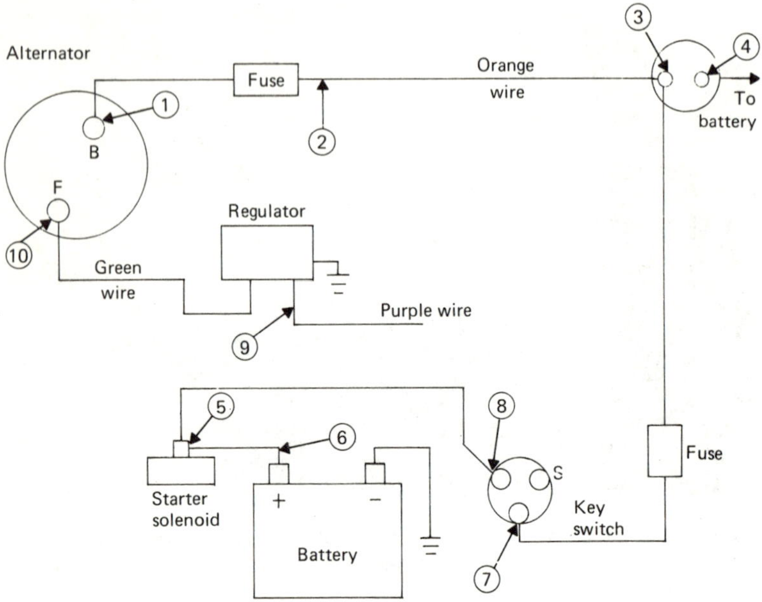

Fig. 4.12 Test-point diagram for checking out the alternator circuit with a 12 volt DC test lamp.

lamp lights the wire, the terminal lug and large fuse (usually located in the orange lead) are OK.

2. Remove the tape or covering from the fuse and connect the free side of the test lamp to both sides of the fuse. If the lamp now lights but did not in the previous test, the fuse is blown and must be replaced.

3. Connect one test lamp lead to the terminal on the back of the ammeter at the operating position on which a heavy red or orange wire is connected. Sometimes this terminal is marked *G*. Connect the other lead to a handy battery ground. If the lamp lights but did not in the previous test at the fuse, then you have a broken or defective connection somewhere in the wire that runs from the alternator fuse at the engine and the ammeter at the operating position.

4. Connect the free lead of the test lamp to the opposite terminal of the ammeter, which is sometimes marked *B*. If the lamp lights now but not in the previous test, you have a defective ammeter.

5. Back on the engine look for the starter solenoid (relay) normally found mounted on top of the starter motor. This is a

major tie point for many electrical supply leads on most small craft, including those with outboard engines. Connect one lead of the test lamp to ground and the second lead to the heavy electrical stud on the solenoid. If the lamp lights here but not in the previous test, you have a broken or defective connection in the usually red wire that runs between the ammeter and the starter solenoid.

6. Connect the test lamp between a good ground and the battery post (*not* the cable clamp). If the test lamp now lights, you have a high resistance connection at the battery terminal and cable clamp. Remove the clamp. Clean and scrape both the clamp and battery post, and reconnect the cable to the battery.

If you still haven't found the trouble, go on to the next series of tests, which are designed to run through the generator/alternator charging wiring.

7. Connect the test lamp between a ground and the terminal on the back of the ignition switch, where a wire runs from the ammeter to the switch (usually orange). On the switch it might be marked *B* or *Bat*. If the test lamp lights here but not in the test No. 3, then the main line fuse is blown and must be replaced. If the lamp still doesn't light, proceed to the next test.

8. Connect the test lamp to a ground or negative return and the back of the key switch (usually a purple wire) and turn the key on, but do *not* start the engine. If the lamp lights, the switch is defective. Before you remove and replace the switch, check the connections for loose and corroded terminals—a common fault.

9. Locate on the engine the voltage regulator with the purple wire going in to it. If there is no terminal, use a common pin or needle to poke through the insulation and connect to the wire conductor. Connect one lead of the test lamp to the pin and the other lead of the lamp to a good ground. If the lamp does not light, you have a bad connection between the on/off switch and the regulator.

10. Connect the test lamp to a ground and to the field terminal (sometimes marked *F*) at the back of the alternator, where a (usually green) wire connects to the terminal. If the test lamp doesn't light here, and did light in the previous test, you have a faulty contact in the regulator. If the regulator is a sealed unit it will have to be replaced. If it is the older electro-mechanical type, remove the cover and clean and burnish the relay contacts of the voltage control relay.

Trouble-shooting the Alternator

The following is a summary of alternator common faults and what to look for:

If the alternator won't charge even into a low battery, look for: a very loose, broken, or worn drive belt, worn brushes or slip rings on the rotor; one or more brushes out of contact with the slip rings; an open rotor circuit either inside or outside the alternator; an open stator circuit either inside or outside the alternator; or blown rectifiers.

If the charge rate as indicated on the ammeter is low or unsteady, look for: loose drive belts; a high resistance connection in the charging circuit wiring or a high resistance in the alternator to engine winding; or an open or poor connection in the stator winding.

If there is a low charge into a low battery, look for: a high resistance in the charging circuit; an incorrect regulator adjustment; a shorted or open rectifier; or a grounded stator winding.

If the charge to the battery is excessive, look for: a maladjusted voltage regulator; voltage regulator contacts welded shut; regulator coil open (burned out); or the regulator itself not properly grounded.

If the alternator is noisy, look for: loose bolts and nuts on the mounting; worn, frayed, and lumpy drive belts; a worn bearing (use the stethoscope method to identify worn bearings); a bent drive pulley or cooling fan; an open or shorted rectifier (causes a whining sound); or an open or shorted stator winding.

If the ammeter changes extravagantly, look for: high resistance in the rotor circuit; a defective regulator; or loose connections in the control or charging circuits.

All or nearly all of the above problems can be prevented by no more than PPM once or twice a season with special attention to the condition and tension of the drive belts. Be sure to keep a spare drive belt and a bit of belt dressing in the boat box.

When the boat is being stored out of the water for the off season, it's a good idea to release the belt tension completely to prevent the belt from stretching and to relieve the pressure on the alternator/generator bearings. If practical, why not remove the unit from the boat, clean it, and store it in a plastic bag until it is needed again.

5
STARTING SYSTEMS

Why do so many small craft skippers find themselves in need of assistance with an engine that fails to start? Most often the problem is traced to a dirty or corroded connection in the starting circuit—the result of a partial or complete lack of preventive maintenance. Furtunately the starting system of a small craft motor is very simple and straightforward. It is essentially what the engineers call a "brute force device," meaning that it uses a large amount of power for a short period of time.

COMPONENTS

Starter Motor

The starter is one of the simplest direct current motors. The copper wire of the armature and field windings is of the heaviest possible gauge to carry the large amount of starting currents used. In most cases four heavy carbon brushes conduct the current from the external connections to the armature by way of the commutator ring. The motor is designed only for intermittent service, because the huge amounts of current used would very quickly overheat the unit. The starter motor will demand more than 100 amperes of current from the battery to start a 200 horsepower engine, which may be the cause of starter failures. To supply that magnitude of current, the system must have near perfect, clean, corrosion-free, tight connections; and the simple maintenance of starting circuit connections is the essence of PPM program.

Fig. 5.1 Example of spiral-type starter shift on outboard motor. Torque of motor sends teeth upward to engage gear teeth of motor flywheel.

The method of momentarily connecting shaft and spur gear of the starter motor with the teeth on the main engine flywheel is somewhat complex and varies with the type and model of engine. The older Bendix spring system seems to have given way to the more modern and less troublesome solenoid operated fork shift and overrunning clutch system. One thing that starting systems seem to have in common is that they all come from the automotive field and even on large outboard engines will be much the same. Electric starter motors themselves, however, have become smaller and more efficient.

A big disadvantage of the helical Bendix spring system is that the sudden forceful slamming of the starter gear into engagement with the gear teeth of the flywheel causes wear and tear and often breaks the large spring (called the Bendix spring), rendering the system useless just when it is needed most. Further, in marine applications it is susceptible to rust and corrosion.

Fig. 5.2 Example of fork-type shifter on starter motor. Solenoid on top of motor both energizes motor and engages teeth of gear to engine.

The shifter fork drive enables the starter motor shaft to engage the teeth of the engine flywheel *before* the starter motor shaft turns. After the main engine is started, the overrunning clutch easily releases the starter motor drive shaft from the flywheel teeth. Thus, most of the mechanical strain of the older system is substantially reduced.

The corrective maintenance of starter motors and starter motor drive units is best left to specialists in this field. However, with but the simplest of *preventive* maintenance the starter motor will outlive the main engine and rarely give serious trouble.

Solenoid Relays

These devices are electromagnetic switches that perform several functions. The *starter solenoid relay* is mounted directly on top of the starter motor. It causes the shifter fork to engage the shaft of the starter motor and sends it forward to engage the teeth of the main engine flywheel. In addition, it carries extremely heavy current from the battery to the starter motor. The average small craft skipper may not know that the starter solenoid also provides extra current to the ignition coil by shorting out the ballast resistor during the starting cycle.

Fig. 5.3 Example of heavy starter solenoid normally mounted on top of starter motor. Note heavy electrical connection stud for battery and other cables.

This heavy duty relay is made up of a heavy wire magnetic winding, a plunger of magnetic metal, very heavy electrical contacts to carry starter motor current, and various other accessories for special applications. The positive main battery cable is connected to the large stud on the solenoid, and this connection is a main distribution point for leads that supply battery current to other electrical systems on the boat. This crucial main electrical connection point, unfortunately, is nearly always very low on the main engine and very susceptible to rust, corrosion, and in extreme cases galvanic action.

If the starter solenoid current demand is so high that it cannot be safely handled by the starter switch, which is mounted up on the dashboard, a *starter assist relay* is added to the system. When the starter switch is turned to the start position, the electric circuit to the coil winding of the relay is energized and the relay closes, since it is nothing more than a heavy current-carrying switch. As current flows to the starter solenoid, it closes and connects the starter motor to the battery.

Neutral Safety Switch and Ballast Bypass Switch

Most small craft with either inboard or outboard engines will have these switches incorporated into the starting circuit, and it is their nature to cause starting problems from time to time. The *neutral safety starting switch* opens the starting circuit when the gear shift lever is in any position other than neutral. Normally it

Fig. 5.4 Starter assist relay is often used as starter solenoid relay draws heavy current.

Fig. 5.5 Example of neutral safety start switch. Unit shown is a microswitch, which must be in closed position or engine will not start. Note adjustable cam. Unit shown is mounted on an outboard engine.

is a microswitch mounted on the transmission close to the shift operating lever. The mounting bolts and electrical connections, of course, are susceptible to rust and corrosion and will become loose from vibration. The Coast Guard has towed a number of small craft whose only problem was an out-of-position neutral safety starting switch.

Another popular mounting position for the microswitch is at the operating position buried under the cover of the gear shift and throttle levers. You should know its location and study the engine manual for the proper alignment of the switch—a simple matter in most cases.

The *ballast resistor bypass switch* may be no more than a set of starting relay contacts, or it may consist of a second set of contacts that operate from the starting relay to shunt out the ballast resistor upon starting and thus provide the ignition coil with greater current and voltage for better starting of cold engines. Once the engine has started and the starting relay is de-energized, the ballast resistor is reconnected into the ignition circuit (primary of the ignition coil) and the current to the coil is slightly reduced during the engine running time. In this manner the bypass switch further protects the points and coil primary from excessive burning. If the ballast resistor burns out, the engine will start but immediately die when the starter key is released to the run position. In an emergency the burned out resistor can be jumped with a short piece of wire, but it should be replaced as soon as you're back at the dock.

Starter Switches

Although not a common source of starting trouble, these switches (usually found at the operating position) have been known to burn out or to fail to complete the starting circuit. The starting section of the switch contacts normally carries a heavy surge of current for a brief period. Corrosion will increase the contact resistance and therefore the heat at those spots. If the contacts become burned and pitted, the starter relay will fail to close. A jumper wire shorted across the terminals on the back of the switch will be enough to get the engine started and the boat back to the dock where the switch can be immediately replaced. Consider the installation of a heavy duty push-button starter switch *separate* from the ignition switch when replacing a burned out starter switch. The better marine push-button starter switches have sealed contacts and are not as vulnerable to corrosion as the others.

Cables and Wiring

A heavy multistranded cable extends to the starter solenoid terminal stud from the battery. On most boats this cable will be the positive pole cable from the battery. Cables, terminal studs,

Fig. 5.6 Typical ballast resistor as might be found on inboard marine engine.

Fig. 5.7 Example of heavy duty marine push-button starter switch.

clamps—everything electrical—cannot suffer the smallest resistance from poor or dirty connections. The cable and its associated connectors must carry 100 amperes or more to the starter motor as it cranks the engine. A dirty spot on a connector that causes as little as one one-hundredth of an ohm of resistance will reduce the starter capacity by almost 20 percent!

The surface of the starter motor where it mounts on the engine must carry the return current back to the negative terminal of the battery. This iron or steel surface is not a good conductor and is very susceptible to the formation of a layer of current-resisting rust. Therefore, mounting bolts must be kept clean and tight. It is always a good practice to add a jumper cable from one of the mounting bolts to the engine ground connector just in case!

Battery

Because this crucial device was discussed in a separate chapter, we shall only emphasize that the battery, like the starter motor and its accessories, will not be a source of either starting problems or embarrassment to you if cared for with a good PPM program.

PPM FOR THE STARTING SYSTEM

An essential part of the PPM program starts before your boating season begins. Take the time to remove and inspect every electrical connection before you install and connect the starting battery. Clean and burnish with either a wire brush or sandpaper every single connection in the starting circuits. There aren't too

Fig. 5.8 Tools and supplies for PPM on the starting system.

many and such a task will take only an hour or two. Reconnect the cleaned contacts, making sure that any flat washers and lock washers used to keep the connections tight have been cleaned also. Spray the connections with WD-40 or any insulating electrical spray or lightly coat them with petroleum jelly or even grease to keep the connections corrosion-free during the boating season. Do not neglect to make sure that the ground connection on the engine block is clean and tight.

Clean and tighten the connections up on the dash or operating position for they—as so many things—are often out of sight and therefore out of mind.

During the season again inspect each connection for tightness and watch for the formation of the green patina of corrosion or simple rust.

SIMPLE TROUBLE-SHOOTING FOR THE STARTING SYSTEM

When, in spite of everything you have done to prevent it, the starter motor does not struggle to turn or even try to start and seems to be completely dead, use this checklist.

1. Test your battery with a hydrometer—it might be truly dead.
2. Look for a loose, corroded, or broken battery cable, and be sure to check both cables end-to-end.
3. Check that the starter motor is tight on its mount—the mounting bolts may have backed off due to excessive vibration.
4. Look for open circuits in the manual starter switch, starter assist relay (if used), and all associated wires and connectors.
5. Check for badly burned contacts inside of the starter relay—removal and replacement seems to be the only cure.
6. Check for an open coil in the solenoid relay by removing all connections and testing with an ohmeter for continuity.

When the starter motor draws current, hums, and growls but will not turn the engine over, use the above tests and in addition look for:

An internal ground in the starter motor winding (remove and repair).

The armature rubbing on field poles inside motor (remove and repair).

A damaged, jammed, or rusty drive shaft in the starter motor (remove and repair).

Water or oil in an engine cylinder causing a hydrostatic lock, which prevents the main engine from cranking (big troubles—call for professional assistance).

There may be times when the starter motor spins merrily but the main engine does not. In this event look for:

Broken, overrunning clutch (remove and repair).

Pinion and shaft rusted, dirty, sticking from old lubricants (remove and repair).

Broken Bendix spring, if used (remove and repair).

Broken linkage or spring in solenoid shifter—pivot pins often wear through and carry away.

Broken teeth on the flywheel starter gear.

Another common symptom of starter circuit failures is the clicking or chattering of the solenoid relay when the starter key is operated, which may be caused by:

A weak or nearly discharged battery.

Dirty or corroded battery and starter motor contacts and cables.

Corroded or burned solenoid relay contacts.

Less common, but known to happen, the starter will not disengage from the main engine after starting. Look for:

Broken solenoid plunger spring.

Faulty manual starter switch that will not open after key release.

Rust, dirt, or corrosion on the starter drive parts.

Most of the above problems are in the catastrophic class and will require professional assistance. However, you can save time and money if you do your own trouble-shooting so that you will *know* what the trouble is and not have to rely on what somebody might *think* it is. The removal of the starter motor and solenoid is not difficult but it is awkward. Rarely is there room enough to get hands, arms, and tools into position to remove the retaining nuts and bolts, which are invariably frozen from rust and corrosion. You may need to remove a large section of decking on an inboard engine boat and may even have to remove an interfering engine cooling manifold. In any event be sure to disconnect the battery entirely before you begin so that violent sparks will not fly if a tool slips and touches the positive lead and the engine block.

On the starter solenoid connecting stud you'll find several electrical leads. Remove and tag them with masking tape so that they will all go back on the same place. Make a note to wire brush or sandpaper all connections and spray with WD-40 before the starter is reconnected.

The starter motor is usually bolted to the engine block with at least two heavy bolts at the coupling flange near the flywheel housing. Be careful not to take off the bolts that hold the starter motor together—a common mistake.

Starter Motor Tests

The following are some simple and quick tests to confirm your suspicions as to the cause of starter motor problems.

1. To confirm that the battery is too weak to turn the starter and is the cause of the problem, turn on one or more lights that are visible from the starter key position and operate the starter key. If the lights dim or go out, the battery is the guilty party and must be charged from the dock or taken out and replaced.

2. To confirm cable connection problems, insert a screwdriver blade between the battery post and the cable clamp as the mate tries to crank the engine. If that cures the problem, remove, clean, and retighten both cable clamps and spray with WD-40.

3. To confirm a neutral safety switch problem, juggle the shift lever to attempt to start the engine. The microswitch may have become loose on its mounting. This switch is normally open except when the shift mechanism is in the *neutral* position.

4. To test the solenoid for proper operation, momentarily short the large terminal on the solenoid to the small terminal with the shaft and blade of a screwdriver. The switch itself must be *on* for this test. Make sure that there are no gasoline fumes present in the bilge when you use this test, as a few sparks will fly as the screwdriver makes contact between the two electrical posts. If the relay operates, the trouble is in the circuit to the starter switch.

Voltage Tests

If you have taken our advice and purchased a pocket-size multimeter, you can perform several simple DC voltage tests to accurately pinpoint the source and nature of a problem.

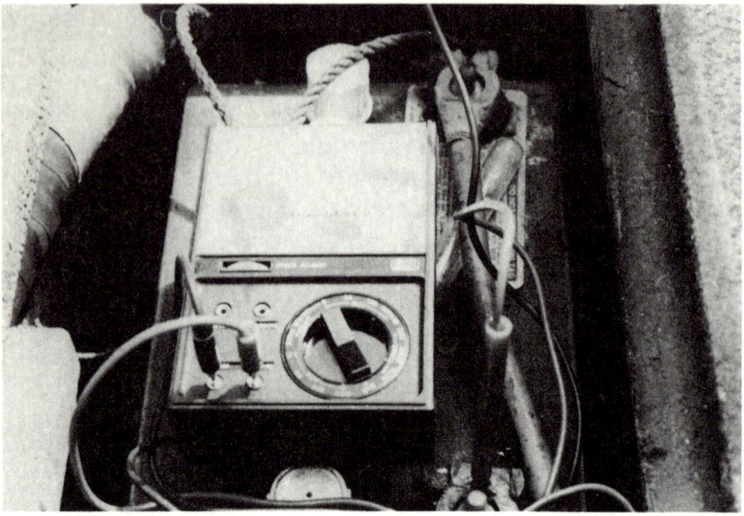

Fig. 5.9 Small meter is connected directly to the battery terminals to test full voltage.

Connect the test leads of the voltmeter to the DC connections and set the meter scale to read less than one volt DC.

1. Connect the positive red lead of the voltmeter to the battery positive post (*not* the cable clamp) and the black lead to a good ground on the engine. Operate the starter key and note the voltage reading on the meter. It must not be more than *0.2 volts*. If it is greater, there is a high resistance connection between the battery and the starter motor connection stud, or the cable itself is too small to carry the large currents.

2. On the starter solenoid itself, connect the red lead of the voltmeter to the large battery cable post and the black lead to the solenoid motor terminal and operate the starter key. Again, if the observed voltage is greater than 0.2 volt, the solenoid contacts have excessive resistance and it may be necessary to remove and either repair or replace the starter solenoid.

3. Connect the red lead of the voltmeter to the *grounded* post of the starter frame and the negative lead to the ground post of the battery and operate the starter switch. The voltage observed here again must not exceed 0.2 volts or there is excessive resistance between the starter motor frame and the engine block.

4. Check the voltage drop in the solenoid switch circuit by connecting the positive red lead of the voltmeter to the large battery stud and the negative lead to the small solenoid switch terminal as you operate the starter switch. The observed voltage here must not exceed 2.4 volts (you may have to switch scales on the meter) for a 12 volt battery. If it is greater, there is excessive resistance in the control circuit which must be assiduously hunted down and eliminated.

5. If in the previous test the voltage did not exceed 2.5 volts and the solenoid still did not pull in, adjust the voltmeter to read up to 15 volts DC, connect the red (positive) probe on the small switch terminal (mounted close to the large battery terminal), and connect the negative lead to a clean good ground on the engine block. Operate the starter key and observe the voltage drop. It must at least read 7.7 volts DC. If less, the solenoid relay winding may be faulty and may need to be replaced.

It should be somewhat obvious from the simple procedures described in this chapter that most starter system problems can be prevented by the application of PPM.

6 SMALL CRAFT WIRING

From time to time the small craft skipper who has adopted the PPM concept will be required to either service or install some new wiring and possibly trace down troubles in the installed wiring. It will be useful if you have a basic knowledge of wiring techniques and symbols on wiring diagrams, and some idea of the wiring of a new accessory. We will begin with a table of wire size. It is of great importance that any wire selected for boat use be of the correct size fot the job it is to do. Especially on small boats, the normal load current is far greater than it would be in a house installation. Small craft voltage, which is nearly always 12 volts of DC coming from the battery or alternator, naturally is far lower than house voltage, and the current is increased to compensate. The greater current requires a greater size wire to safely accommodate it or there will be a significant rise in temperature in the wire as the current flows.

TABLE 6.1
Minimum American Wire Gauge (AWG for 10% Voltage Drop (12 volts)

Current (amperes)	Length of Wire 10	15	20	25	30	35	40
5	14	14	14	14	14	14	14
10	14	14	14	12	12	12	10
15	14	14	12	10	10	10	8
20	12	12	10	10	8	8	

INSULATION

An electric current will flow through many of the materials on a boat. A surface film of dampness will conduct a significant amount of current. Leakage current paths are often a serious problem on small craft for they will tend to drain down the battery for no apparent reason. Thus, it is necessary not only to choose the correct size wire but also to select the proper type of insulation for the job the wire must do. Table 6.2 is a guide for selecting insulation.

TABLE 6.2
Wire Insulation

Use	Type Designation	Maximum Temperature	Description
General	THW	167°F	Thermoplastic, heat and moisture resistant.
	RHW	167°F	Rubber, heat and moisture resistant with oil-resistant neoprene jacket
General, except machinery spaces	TW	140°F	Thermoplastic, moisture resistant, flame retardant
	RW	140°F	Rubber, moisture resistant, with oil resistant neoprene jacket.
	RW-RH	140°F	Rubber, heat, and moisture resistant with oil-resistant jacket.

COLOR CODING

Having chosen the proper size wire and the best insulation for the job the next step is to carefully consider the color of the insulation. Nothing is more frustrating when trouble-shooting an electrical problem than staring at a large bundle of wires all of different colors and wondering what goes to where. Boat wiring color coding is yet to be standardized, but the American Boat and Yacht Council is working on it. If the maker of your boat followed the recommended color code system you will find that wire colors will be run according to Table 6.3.

WIRING DIAGRAM SYMBOLS

Most small craft skippers are constantly improving the safety and navigation systems aboard their boats and include many of the comfort and convenience items that require some form of

TABLE 6.3
Color coding of wires

Color	Circuit/Unit	Use
Yellow with red stripe	Starting Circuit	Starting switch to solenoid
Yellow	Generator or alternator field control circuit	From regulator terminal to Generator/Alternator field terminal
	Bilge blower fans	Fuse or switch to blower
Dark grey	Navigation lights	Fuse or switch to lights
	Tachometer	Tachometer Sender to gauge.
Brown	Generator armature	Armature to regulator
	Pumps	Fuse or switch to pumps
Orange	Accessory feed	Ammeter to alternator or generator output and accessory fuses or switches
	Accessory common feed	Distribution panel
Purple	Ignition	Ignition switch to coil and instruments
Dark blue	Cabin and instrument lights	Fuse or switch to lights
Light blue	Oil pressure	Oil pressure sender to gauge
Tan	Water temperature (engine)	Temperature sender to gauge
Pink	Fuel gauge	Fuel gauge sender to gauge

wiring. Most of the items purchased for installation on small craft will have packed with them small and sometimes complex installation wiring diagrams. If you are familiar with the symbols used and their meanings, these diagrams can aid in proper and safe installation. Table 6.4 explains many of the diagrammatic symbols used on wiring diagrams.

TABLE 6.4
Wiring Diagram Symbols

Symbol	Meaning
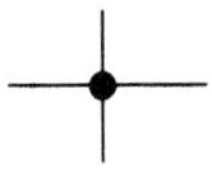	Connection (Wires are electrically joined together.)
	No connection (Wires cross but are insulated one from the other.)
 or 	Ground or common negative return. (Observe this symbol with care. It often has different meanings to different draftsmen.)
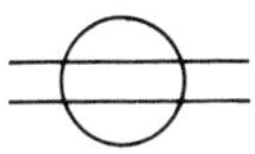	Two-conductor cable (Note: Shield is often grounded at each end.)
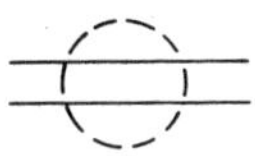	Shielded two-conductor cable (Note: Shield is often grounded at each end.)
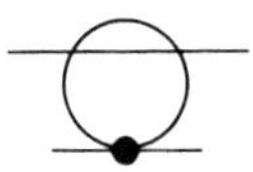	Coaxial cable (Used in radio-radar or other radio frequency type of installation.)
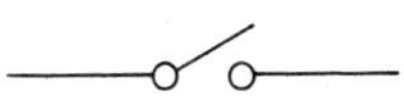	Single pole -double throw switch, such as a running light switch.
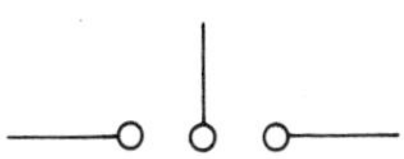	Single pole—double throw switch, such as battery transfer switch.
	Push-button circuit breaker
	Fuse
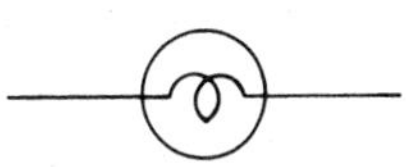	Lamp
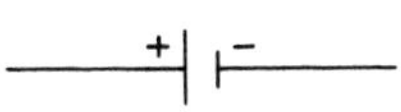	Single battery cell (The long vertical line is always the positive terminal.)
	Battery
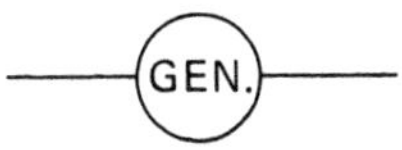	Generator or alternator, depending on how labeled.
	Solid state rectifier (The positive side is in back of the arrow.)

Symbol	Meaning
	Resistor
	Capacitor (Observe polarity if marked.)
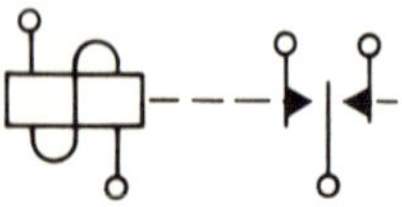	Relay coil with contacts
	Transistor, PNP Type

MATERIALS AND TECHNIQUES

Crimp Lugs

These devices have taken many of the trials and troubles out of wiring for the do-it-yourself skipper. It is important to note, however, that the "solderless" feature must be discarded if they are to be used on small craft. On land the lug performs flawlessly without the addition of a drop of solder after the lug has been crimped to the wire. On the water, however, air and salt corrosion attacks the metal of the crimp lug (often coated with zinc) where it makes contact with the copper wire, and very soon a build-up of current resisting chemicals is formed. The resultant high resistance connection, which usually cannot be seen, causes a mysterious drop in voltage and current. Two steps will, in all but the most stubborn cases, eliminate this problem. First, carefully tin the strands of wire before insertion into the crimp lug. Second, add a drop or two of solder to the intersect of the wire and lug after it has been crimped. The tinning is done with rosin core solder and either a small soldering iron or a very small gas torch if the job must be done on the boat. Take all the usual precautions before using an open-flame torch: ventilate the boat; make sure no gasoline fumes are present; and place a fire extinguisher next to the work.

A crimp lug can be found for just about every job. For small craft wiring you must use a ring type lug that will not work loose and fall from the terminal, as a hook or spade type lug might.

Choose the correct shank size for the wire and the correct ring size for the terminal it is to be affixed to. Most crimp lugs have plastic insulation already on the shank of the lug. For small craft wiring, however, the uninsulated lug may be better because of the need to add solder after the lug has been crimped. Before affixing the lug to the end of the wire, cut and slide a small length of shrink tubing insulation up the wire and out of the way for the moment. Then crimp the lug to the wire and add the solder. Slide the tubing up over the shank of the crimp lug, add a bit of heat to the tubing, and watch it shrink and tightly conform to the shape of the lug shank. You have created a trouble-free joint in which the connections are insulated from contact with other current-carrying circuits and the wire and lug interface is virtually airtight.

Splices

As a general rule splices in rope, wire, or even cooling water hoses are an abomination on small craft and are to be avoided whenever possible. However, there are times when a splice is necessary, such as when installing an accessory that is delivered with a pig tail set of wires to which you must connect the feed and ground wires.

The easiest and least troublesome splice is made with crimp connectors designed for that purpose and the same technique as used for stud terminal crimp lugs. In an emergency, you can use one of the illustrated splicing methods. Follow one basic rule: Never splice a smaller wire to a larger wire. It is much better, if there is no other wire available, to join an oversized wire to a smaller wire.

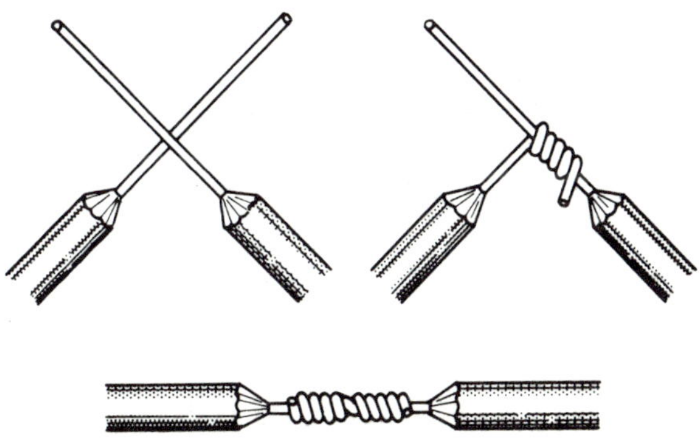

Fig. 6.1 The Western Union splice.

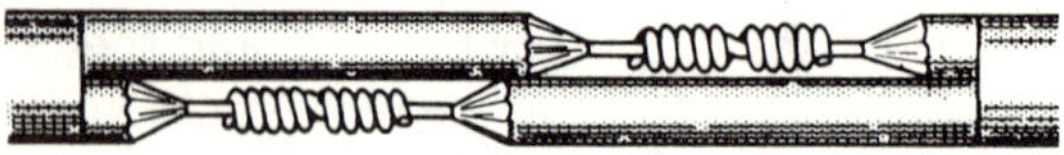

Fig. 6.2 The staggered splice avoids a bulky taped joint in the two wires.

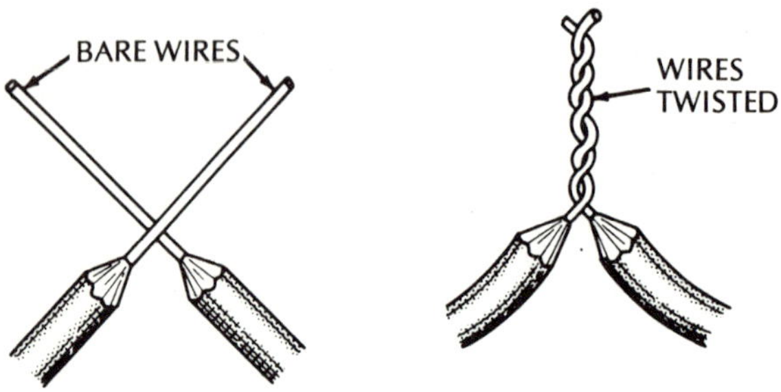

Fig. 6.3 A simple rat-tail joint for a temporary splice.

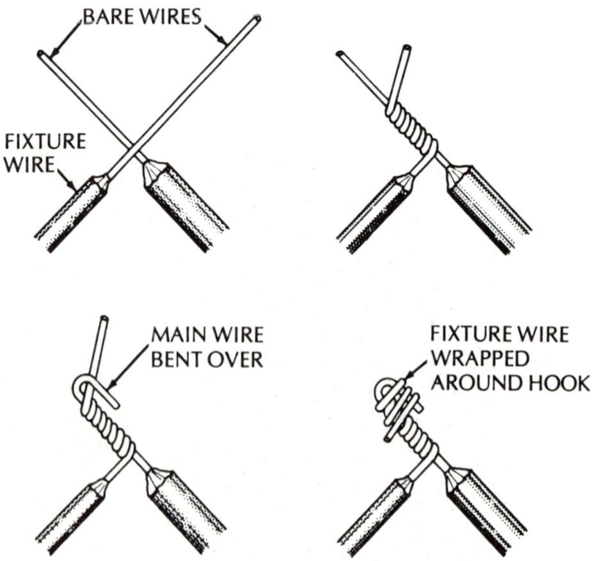

Fig. 6.4 A fixture joint might be used to connect a cabin lighting fixture to the main lighting feed.

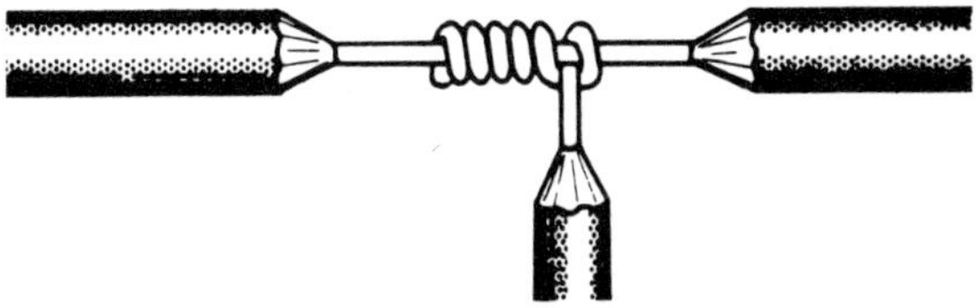

Fig. 6.5 A knotted tap joint might join several lamps to the feed wire. It is not necessary to cut the main wire but only strip a small section of the insulation and add the tap joint as shown.

When removing the insulation from wire for these types of splices, a different technique is used that is similar to sharpening a pencil and avoids nicking the wire. In the illustrations, solid wire is shown for simplicity, but you should *always* use stranded wire on small craft. After the insulation is removed, the bare wire ends should be scraped bright with the edge of a knife blade or rubbed clean with fine sandpaper.

Cable and wire clamps

Proper, safe, and trouble-free wiring techniques require the use of clamps to hold the wire runs in place. The function of the clamps is to prevent excessive movement of the wire and thus prevent chafing through the insulation with the resulting short circuits. On small craft there is substantially more vibration than in a car or at home, and the wire and cables must be firmly held in place. In addition, cable clamps make a neat and easily traceable installation.

Unfortunately the most common and easily obtainable types of cable/wire clamps are fabricated of metal and are almost totally unsuitable for use aboard small craft where they will rust and corrode. The best clamps for boat use are nylon or plastic, and they can be found at marine hardware stores and electrical supply stores. Their soft and smooth surfaces wiil hold wire or cable firmly in place without danger of chafing. Normally they will require either wood screws or self-tapping screws which are not susceptible to rust or corrosion. Bronze, chromed, or stainless steel screws are the best choices.

It is rarely convenient to use cable clamps when running wire down the side of a fiberglass boat, and it may be necessary to bond small blocks to provide an attaching point. Blocks can be set in epoxy resin glue under floorboards or gunwales where they will be out of sight.

Fig. 6.6 A metal wire clamp should *never* be used on boats!

Conduit

The use of PVC piping as conduit will keep wiring together and make a neat, trouble-free installation. All PVC joints are easily bonded with a glue provided for that purpose. This plastic piping can be run through bilge areas and under engines; and the wiring that is fed through the piping will not be subjected to oily wastes and bilge water. Because the piping is smooth and slippery inside, it is easy to fish wire through if a new line must be run.

Heavy flexible plastic tubing is also available from electrical supply stores and can be used for the same purpose. Wire and cable are bundled together and the tube is pulled over the handle. The entire cable bundle can then be clamped into place with plastic cable straps.

Grommets

Whenever it is necessary to feed a wire or cable through a hull member such as a bulkhead or floor stringer, it is necessary to use a grommet to prevent the chafing of the wire insulation against the drilled and often sharp-edged hole. Soft rubber grommets come in a variety of sizes and can be found in electrical supply stores and marine hardware stores.

Another protective technique is to fill the hole through which the wires have been passed with room temperature vulcanizing rubber (RTV) so that it extrudes on both sides of the hole. This odd and wonderful material, which comes in semi-liquid form in a tube, soon sets to rubber and then may be trimmed to provide a perfectly shaped grommet that is water-tight and holds the wire or cable firmly in place. For further protection, wire clamps should be installed on each side of the feed-through hole.

Cable Straps

When a number of wires must be bundled together, nylon cable straps will do the job. They are self-locking and very tough, will not rust or corrode, and may be easily cut and replaced. Use of these straps will keep large bundles of wire neat and out of the way (dressed) and prevent excessive movement and chafe. Cables that are bundled and strapped are easily clamped into place.

Wiring Tools

The following tools are suggested for doing either preventive or corrective wiring maintenance on small craft. Not all are needed and often one tool can do the job of several others.

Wire cutters, heavy snips, linemans pliers, or diagonal cutters.

Wire insulation strippers (there are many different types available).

An electricians knife, a spin-tight socket set with insulated handles, a pair of needle-nosed pliers or round-nosed pliers for shaping wire loops.

A soldering torch (either gas, alcohol or a 12 volt DC electric will do most soldering jobs on small craft).

A multipurpose terminal crimping tool (often available in a kit with various sizes and types of crimp lugs; also cuts and strips insulation from wire).

Miscellaneous Materials for Wiring

These materials are suggested to make a wiring job easier and trouble free:

Rosin core solder (never the acid type for small craft).

Plastic electricians insulating tape, which has many other uses on small craft.

Hook-up wire (small spools of various sizes and colors).

An assortment of crimp terminal lugs (which store in plastic 35 mm film cans).

Insulating spray (WD-40 or similar).

Vaseline (tube).

A small brass/bronze wire brush and sandpaper for burnishing contacts.

Miscellaneous small fasteners (nuts, bolts, lock and flat washers).

Plastic cable clamps, nylon cable straps, grommets, and a tube of RTV rubber.

Spare fuses, lamp bulbs, switches, and lamp sockets.

FUSE BLOCK AND DISTRIBUTION POINTS

The average small craft skipper might need to install a new central electrical distribution point, either because there was none when he acquired the boat or the wiring is now a confusing mess, or because the present fuse panel has an excessive number of circuits connected to it and it should be replaced.

It is important to note that this task is far more tedious than it is technically difficult. If you work slowly but thoroughly, you'll end up with a trouble-free electrical distribution point in which every circuit is known to you.

The first step is to study the present electrical load to determine the required number of separately fused circuits. Except for general lighting, each circuit should be fed from one fused point that is properly fused for the expected current drain. For example, you will require one circuit and one fuse for each of the following: the engine compartment air blower; the bilge pump(s); cabin and cuddy general lighting; the running lights; each unit of electronic equipment; the spotlight or docking lights; and instrument panel lights.

The next step is to examine the present wiring to see if the wire is adequate for the current to be conducted and if the color is correct for the circuit (see Tables 6.1 and 6.3). While you're about it, check the insulation for chafes and kinks, and the general condition of the terminal lugs. Correctly rewire any substandard circuits, eliminating as many splices as possible. Tag

Fig. 6.7 A fused and switched circuit panel is useful on small craft.

the ends of the wire to identify the circuits before you select the fuse panel central distribution board.

To determine the total current drain that the new fuse panel must provide, add to your present requirements the estimated current drain of electrical equipment you may install in the future. Then add a safety factor of 10 percent. If you have a a present load of ten separate circuits, your choice of a circuit board should have at least five additional circuits for use at a later date. The present current load will give you an idea of what to expect for any new circuits. To determine the current required by a given circuit, for example a lighting circuit, add up all the wattages of all lamps on the circuit and divide the total power in watts by the voltage (usually 12 VDC) to obtain the current in amperes. If the only information you have is the number printed on the base of the lamp bulbs, take down the number and ask your dealer what current this number bulb draws. Current drain, or the power consumed for motors and electronic equipment, is usually stamped on the outer case or is provided with the installation instructions.

Next, run a new feed wire set from the main boat battery terminals up the side of the boat to the distribution panel. Select a rugged, two conductor, well insulated, stranded wire cable, sized for the current drain. Install nylon cable straps and plastic cable clamps every few feet along the run. It is best to connect this feed wire directly to the cable clamp or battery terminals and *not* tack it onto the starting motor solenoid terminal. Once this feed cable has been run, disconnect it from the battery terminals until the balance of the wiring has been completed.

The best type of distribution panel is one that mounts on or very near to the operating control position and includes on/off switches and twist plugs in the fuse holders. This type of installation may require the cutting of an adequate hole in the wood panel at the operating position so you can control the boat's electrical distribution without leaving the piloting seat. Simpler fuse panels require the addition of switches at other locations, and this type of system is common to small craft. In any event, try to locate the panel so that circuits can be checked and fuses replaced with relative ease.

Install the panel, using adequate hardware that will not corrode and cause trouble later. Carefully connect the main feed wire using heavy crimp lugs, insulated and soldered to the buss bars of the panel. The feed wire conductors should be connected black to positive (hot) and white to negative (ground or negative return). Tagging or other identification of this circuit should not be necessary. When attaching the terminal lugs be sure to use flat washers and lock washers on the ring terminals and studs. Install a wire clamp close to the main feed wire and the buss bars.

Next, install each of the circuits, one by one, to the buss bars at the fuse terminals (leave the fuses out of the circuit for the time being). Generally all of the negative return wires can be connected to the negative return buss at any convenient stud. As each circuit is installed and connected, turn its switch on and momentarily touch the main feed cable to the battery terminals.

It is helpful to have another person work at the battery end as you install the fuse and check the circuit for proper operation. After each circuit is checked, remove the battery terminals until all circuits are completed. This precaution will prevent potentially dangerous arcs if a tool should slip. Carefully and neatly dress the wires and use nylon cable straps to hold them in place. When the circuits have been installed and connected, insert the fuses and again test each circuit for proper operation. A really classy installation would include a small lamp and switch over the fuse panel to provide lighting during trouble-shooting.

In addition to following the color code table, using tags, or placing plastic printing tape under each fuse will help to quickly identify a given circuit.

CENTRAL GROUND POINT

Everything electrical in the boat must be electrically tied to a central ground point, which is *not* the negative post of the battery or the bolt or nut that grounds the battery to the engine. On some boats there is a large heavy copper or brass plate centrally located to provide a common electrical tie point for all negative returns.

It is always convenient to run wire to this central ground point. The best installations employ a ground strap or bonding conductor, which is simply a tube of copper or a flat ribbon of copper or aluminum that runs unbroken from the stem to the stern and is bolted or screwed to wood frames along the top of the keel area. It is important to understand that this heavy ground strap does *not* carry any load current. The normal positive and negative leads or cables from the battery carry *all* of the current. The central ground strap serves to connect, electrically, all of the equipment and systems in the boat so that they are all at the same electrical potential or voltage. Galvanic action cannot occur under these conditions. A separate wire bonds the metal chassis of the radiotelephone to the central ground strap. A 12 volt anchor windlass on the bow has a separate wire that bonds all metal parts of the windlass to the ground strap. Even the lead tray of the battery box (assuming you have one) is bonded to the central ground strap.

When a watertight bulkhead prevents the ground strap from running unbroken fore and aft, the strap is connected with a through-bolt from one side of the bulkhead to the other. The bolt hole is drilled large enough for the shank of the bolt, and a waterproof bedding compound is placed under the ends of the straps where they are joined with the bolt, nut, and washers. Fasteners used to hold the ground strap in place along the keel must be of more noble metal than the ground strap so that they will not corrode. Copper, silicon, bronze, Iconel, or 18-8 type 304 passive stainless steel fasteners are preferred.

BONDING JUMPERS

The wires used to connect equipment and systems to the central ground strap need not be insulated unless they pass close to current-carrying wires. At least No. 8 wire should be used and kept as short as possible. These jumpers form the "ribs" of the system. A bonding jumper run from a brush pick-up on the drive shaft to the central ground strap will help to eliminate static noise from the radio and prevent galvanic action between the propeller, shaft, and rudder. Rudder posts, strut mounting bolts, and outdrives should have a jumper connecting them to the central ground strap. Fuel and water tanks, metal fuel lines, and deck fill fittings should be bonded to the central ground strap. The *entire fuel system* should be carefully bonded to the central ground to prevent fires and explosions caused by static discharges and lightning side flashes.

The ground strap should be connected to the same place on the main engine(s) as the negative cable of the battery. Large metal cabinets and electrical/electronic equipment must have a bonding jumper connected to the grounding strap. Fuse boxes, metal circuit breaker panels, electric refrigerators, and stoves all get the same treatment.

Fortunately for the do-it-yourself skipper, installing the central ground strap and connecting the bonding jumpers is not an urgent, all-at-once job. Get the ground strap in and tightly fixed to the center line of the boat. Then, when you have the time, connect the bonding jumpers—but keep at it until the job is done.

The central ground strap and bonding jumpers will reduce or eliminate galvanic action, and electronic noise from the communications/navigation equipment. Equally as important, they will reduce the hazard of electrical shock especially to persons aboard boats with dock power and 115 volt AC generators, and will go far toward providing lightning-strike protection.

PPM FOR LIGHTING CIRCUITS

It is a rare boat indeed, other than a skiff or a dinghy, that does not come already equipped with proper and legal running lights. All boats that may be operated either at night or in reduced visibility *must* be properly equipped with running lights. What is "proper and legal" depends on the type of boat (sail or power), its length, and whether it is to be operated in waters that the

Coast Guard has defined as *inland* or *international*. Production boats often are delivered with running lights installed under the rules of international lighting and can be operated in any waters. Boats presently equipped with lights for operation on inland waters, which include waters along the sea coast, cannot be operated legally outside of those waters at night.

Most small craft skippers will have few problems with running lights and what problems do occur are caused by disuse and corrosion in the lamp bulb sockets. Although the running lights are designed for exposure to salt spray and rain and are said to be weatherproof, oxygen and moisture will get inside of the fixtures and create corrosion between the base of the bulb and the socket. The bulbs nearly always have a bayonet-type base, and contact is maintained by a steel spring that presses the electrical connections of the socket and bulb together. The spring often rusts and adds to the problem of corrosion.

PPM will reduce and very nearly eliminate lighting circuit problems. Early in the beginning of the season the running light fixtures should be disassembled and inspected for signs of corrosion. Sockets and lamp bases should be cleaned and burnished with fine sandpaper and then given a liberal dose of WD-40 insulation spray or its equivalent. If the spring is accessible, it too should be sprayed. Wiring must be checked for chafes and kinks all the way back to the fuse panel and on/off switch.

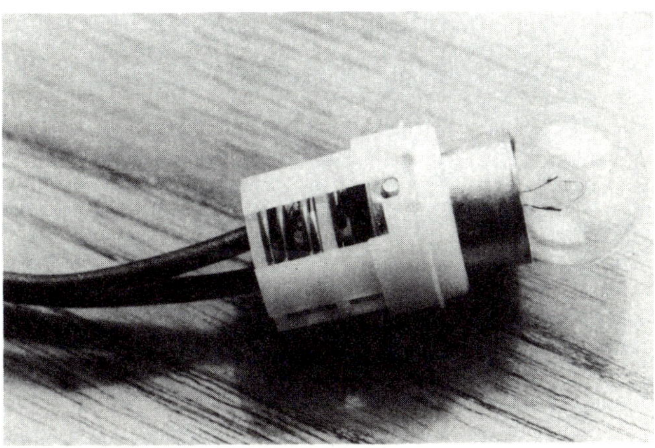

Fig. 6.8 This plastic lighting socket for 12 volts DC is excellent, but the metal spring will rust unless sprayed and watched.

If the metal ends of any glass fuses are given a light burnishing along with the fuse holder spring clips and then sprayed, they will ordinarily last all season without corrosion troubles and the resultant high resistance connections.

As a rule lighting control switches are of more rugged construction than those found on automobiles. Still, they have been known to burn out the internal switch contacts—again due to corrosion of the metal switch contact faces. The resistance causes heat when the switch is on and eventually the contact faces will burn to the point where the switch may become a permanently open circuit. Some switches can be disassembled enough to expose the actual switch face contacts so they can be cleaned with a small pen knife and then sprayed and reassembled. Do not neglect to clean the wire stud contacts on the back of the switch and give them a light spraying after reconnecting the fixture leads.

Spot or search lights, spreader lights, and bow lights for the most part use modern seal beam lamps of very high intensity light capacity. While the lamps themselves do not often fail, the sockets, bases and wiring are vulnerable to corrosion and high resistance connections and should be treated with the same PPM techniques.

Although not required by law or regulation, instrument or panel lights and compass lights are essential to safe running at night. In most cases instrument panel lights are small low

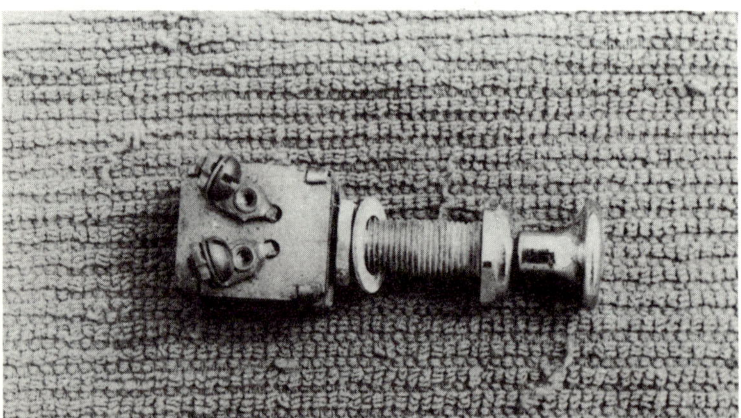

Fig. 6.9 Example of a marine service 12 volt lighting switch (pull type).

Fig. 6.10 Example of an instrument lighting socket for a tachometer.

intensity lamps that plug into the back of the instrument to be observed. Customarily, a single wire is led over to the running light switch so that the instrument lights are energized at the same time the running lights are lighted. The ground or negative return to the battery is provided by connecting the metal case of the instrument to the engine block. Preventing loose and corroded connections will eliminate most of the problems with these circuits. Always replace these lamps with very low light level lamps so that your night vision will not be upset from looking at your instruments. The compass lamp should also be coated with red dye of some sort.

DUAL BATTERY SYSTEMS

Cabin or bow cuddy lights are often left on long after they are needed and run down the starting battery, perhaps when at anchor. To prevent this problem, you may want to install a dual battery system. Other than the normal and required safety equipment, few things make better sense on small craft!

With a dual battery system properly installed there is always a fresh fully charged battery available to start your engine. Most of the larger outboards and all of the inboard engines can no longer be manually started when the battery has apparently failed. Even when your PPM program has been faithfully carried out, things can still go wrong. For example, you might have anchored in your favorite fishing spot for an afternoon of drowning a few sea worms. At the end of the day you get ready to up anchor and operate the starter key to start position. You get a sinking feeling in your stomach when you hear a clicking sound from the starting relay but no reassuring whir of the starting motor. Either the bow cuddy lights or the depth sounder has been running all afternoon, and you've used the radio telephone. In any event there's not enough energy left in the starting battery to turn the engine over. Hopefully, there is another boat nearby to give you a tow or there is enough energy left in the battery to get out an assistance call to the Coast Guard—at which point you become another statistic in their files!

The added cost of dual battery systems is more than offset by the safety and comfort of knowing that you always have that extra battery for starting. Once the engine is rolling, the alternator will take over the electrical load and almost immediately recharge the battery.

Bringing a second battery aboard and merely connecting it in parallel with the original battery does not solve the problem. What is required is to have two completely isolated batteries that can be charged from the same alternator but switched into and out of the circuit as desired. The second battery does not have to be a high quality marine service battery; a good used automotive battery with a lot of life left in it or a low cost battery with the capacity to start your engine will suffice. A second set of battery cables with clamps and heavy lug terminals will also be required (more on these later). The third element is a vaporproof, high capacity, make-before-break dual battery transfer switch, which can be purchased in most marine hardware stores for about $20. Most of these switches will have several positions: for example, one position for the starboard battery only; a second for the port battery only; a third for batteries connected in parallel; and a fourth position for off, or both batteries completely disconnected from all circuits. This last switch position prevents stray currents from draining down the batteries when the boat is left for long periods of time without the engine turning over. In addition it will prevent destructive galvanic action in the battery circuits.

Here's how the system might be used on a cruise. You start the engine with the switch in Position No. 1 (starboard battery starts the engine), and leave the switch in this position until

Fig. 6.11 A dual battery installation with transfer switch mounted on inner transom area of boat.

Fig. 6.12 A dual battery system transfer switch. Note the warning! A better switch is the make-before-break type.

your ammeter indicates that the starboard battery has been restored to full charge by the alternator. At this point you operate the switch to Position No. 2, which places the port battery into the circuit and at the same time places the starboard battery in a disconnected reserve. The port battery is quickly charged by the alternator, and you can remain in that switch position or return to Position No. 1. Now you have two fully charged batteries ready to go to work. Should anything run down the battery that is on line, the second stands ready to start the engine and get the alternator on the line.

A good quality transfer switch will be of the make-before-break type; that is, as the switch is operated from one position to another, contact will be made with the second position before the contacts leave the first. Thus the diodes of the alternator will never look into an open circuit where high voltages are present that can easily burn them out.

Installation of a Dual Battery System

The space available in and the needs of small craft vary substantially. Thus it is not possible to outline specific instructions for the installation of a dual battery system. However, several general rules can be given, and most skippers will not find the installation difficult. As always first and foremost, the installation should be carefully planned. Very likely the task will take much less than a full day, but you would be wise to allot that much time. If possible, the second battery should be installed on the opposite side of the boat from the original battery and as close as practical to the engine starting circuits to keep cable lengths short. Plan to construct a second battery box with lead lining and rugged hold-down clamps or straps that will not rot, rust, or corrode. Get the second battery installed in its permanent position before you go to the next step, for it will save a lot of time and trouble later.

The next step is to carefully remove both the positive and negative battery cables from the present starting battery and set them aside for the moment.

The third step (assuming you have purchased a battery transfer switch) is to unpack the switch, admire it for a moment, and set it aside. Normally, there will be an installation instruction sheet packed with the switch. Read and absorb the information contained on the sheet. When you are satisfied that you understand the installation requirements and how the switch works, you can begin to consider possible locations for the switch. Keep these factors in mind: sufficient space is needed behind the switch for the heavy battery cables to connect, and it may be necessary to use wood or plastic mounting blocks. The switch should be located where a second set of cables for the original battery will not be necessary. Also, the switch should be as high out of the bilge area as possible and quickly accessible.

Loosely mount the switch and test for the preceding requirements. Rerun the positive lead of the starboard battery to the transfer switch, adding plastic cable clamps every few inches. If it is possible to shorten this cable by cutting off the old terminal lug and soldering on a new one, do so. Loosely connect the end of the cable to the stud on the switch following the instructions packed with the switch. Run a new lead from the switch to the starting solenoid, again following the switch instructions. Reconnect the negative battery cable to the engine, tighten up the new cables from the battery to the switch and the starting solenoid, and test to see that the engine will start from the No. 1 position. If all is well, again remove the positive lead from the battery to prevent accidental tool shorts and get ready to wire in the second battery. Using a scrap of light but stiff wire, measure the length of the cable run from the positive pole of the second battery to the switch terminal for Position No. 2. Also measure the length of the cable needed to ground the second battery to the same stud as the original battery. Purchase the needed cable, clamps for the battery, and stud terminals for the switch end of the cable.

In making up these cables take pains to do a neat job of attaching the battery cable clamps and the studs. If possible, solder both ends to clamps and terminals and cover with shrink tubing or plastic electricians tape. Attach the new cables to the stud on the switch at Position No. 2, and attach the new negative battery cable to the same stud on the engine as the original battery negative terminal is now connected, making sure that there is a flat washer and lock washer under the hold-down nut. Give this connection a good spray of WD-40 or the equivalent. It will not be necessary to run a second positive lead from the switch to the starter solenoid, for the old lead (or new one if necessary) is now common to both batteries. Again attach the cable clamps to the second battery, set the switch to Position No. 2, and test the starting circuit to make sure all is well.

It but remains to test Position No. 3 (both batteries in parallel) and Position No. 4 (off or no battery connection) and the installation is complete.

Some skippers find it difficult to remember which battery was used last and which battery should be used to start the engine. However, the switch can be operated to either position to quickly start the engine and the ammeter watched for reaction. The battery used for starting quickly comes up to full charge, at which point the switch is thrown to the second battery and left there for the remainder of the cruise or until the first battery is needed again.

Electronic Dual Battery Isolators

It is not surprising that electronic manufacturers have come up with a better way to control and distribute electric power from a dual battery installation on small craft. The device, called a dual battery isolator (DBI), when wired into the electric power circuits, eliminates the need for any switch and switching. What may be a surprise is that one of these electronic units is less costly than the high capacity vaporproof switch it will replace. The unit is manufactured by the Coloroid Company, 10 Delmar Road, Medway, Massachusetts 02053, and cost less than $15. Inside the unit are two heavy current capacity solid state diodes connected back to back. These diodes are mounted inside an aluminum heat sink which assures complete cooling when in operation and protection from atmosphere. Three connecting terminals appear at the top, making it simpler to wire this unit than the battery transfer switch.

To install this unit select a mounting area near a good stream of cooling air in the engine compartment and as close as possible to the two batteries. The output wire from the alternator is removed from the stud (usually on the starter solenoid) and reconnected to the center stud of the DBI. The positive cable from one battery is connected to the *A* stud of the DBI and the second positive cable from the second battery is connected to the *B* stud. The cable from the starter motor solenoid is then connected to either *A* or *B*. All other electrical loads (wires usually found connected to the starter solenoid stud) are now connected to the DBI terminal except the one used for the starter motor. Electrically, battery *A* and battery *B* and their electrical loads are now completely isolated from each other. One battery is always ready and freshly charged to start the engine while the second battery provides all the boat's electrical needs and normally is lightly loaded. The alternator will charge both batteries at the same time if needed without throwing a switch into a specific position or will charge only the battery that requires it. From now on the skipper need not concern himself with battery problems and may concentrate on PPM of the terminals and cabling.

The electronic DBI has yet another effective use. A number of small craft (even those of 25 feet and less) are installing 115

Fig. 6.13 The dual battery isolator will replace the transfer switch.

volt AC dock power to provide for small appliances when tied up to the dock. Among the more useful appliances is a battery charger that is installed in the engine compartment and used to maintain the charge of the batteries when at the dock. Although inexpensive trickle chargers can be used for this purpose, they may damage batteries when left on charge for long periods of time. The better unit is an automatic device that keeps the battery voltage up to full charge and then shuts itself off. A typical unit is the Constavolt® made by the Surrette Battery Company of Salem, Massachusetts.

The DBI come into play in the use of a dock power battery charger and a dual battery system by eliminating the need to switch the charger from one battery to another and the need to remember which battery is fully charged. The output of the battery charger is connected to the center stud of the DBI while battery *A* and *B* are connected to the two outboard studs. In this manner both batteries will be maintained at full charge at all times the boat is on dock power. Underway the alternator output is also connected to the center stud of the DBI and takes over the charging function.

7 IGNITION SYSTEMS

The small craft skipper finds that much of his corrective maintenance and a good portion of his PPM program is devoted to the care of the electrical ignition system on the engine. In Book 2 of this series, *Engine Care and Repair,* trouble-shooting and corrective and preventive maintenance were covered in detail. The following paragraphs cover the same subject but from an electrical viewpoint. We will trace the ignition system of a gasoline engine from the ignition switch to the spark plugs and explain the function of each part, what happens when that part fails or symptoms that it will give you, and how to do your own tune-ups.

IGNITION SYSTEM COMPONENTS

Ignition Switch

For most small craft the ignition switch is a two-position switch. In the *off* position, no current can get to the ignition system. In the *on* position, the battery is connected through the switch to the point set and ignition coil and the engine is ready to start. *Start* is usually a spring loaded position for the momentary connection needed to get the starter motor to turn the engine. As the switch key is released the spring loaded switch returns to the on position.

Switch problems

That ignition switches on small craft are similar or identical to those on automobiles is a problem. The automobile atmosphere is more or less friendly to a switch, and the switch rarely causes

Fig. 7.1 Example of a lock-type ignition switch. Note the three connections at base.

trouble. On a boat, however, switch failure is somewhat more common. Vibration loosens the connections, and corrosion forms on terminals and wire lugs and often on the contacts inside the switch body. The contacts become heated and break or become high resistance connections, and a weak spark at the coil makes for hard starting. If the start side of the switch is involved, the starter solenoid and motor may operate intermittently or not at all.

Test the switch by using a 12 volt test lamp, making sure that voltage is apparent at each of the terminal lugs on the back of the switch. In an emergency you can start the engine by jumping with a short piece of wire on the *run* position terminal and shorting between the *start* lug and the *bat* lug with a screwdriver. If possible, carry a spare switch in the boat box.

Wire to Coil

On most small craft a wire will run from the ignition switch down the side of the boat to the engine, where it normally terminates on the primary side of the ignition coil. Current flows from the battery through the switch, the wire, and the primary of the coil. Usually this wire is ordinary stranded copper, but in some cases it might be a special alloy resistance wire whose purpose and function is discussed in the next paragraph. A few problems crop up and are always the same—loose connections or corroded terminals that can be prevented by routine inspection, cleaning, and spraying with WD-40 or the equivalent.

Ballast Resistor

This low resistance device (normally 1 ohm when cold) is connected between the ignition switch and the coil primary stud and limits the current flowing to the coil. When the key is turned to the start position, the ballast resistor offers very little resistance to the flow of current to the coil; and the coil, therefore, puts out a greater spark voltage for starting. As the engine warms up, so does the ballast resistor (due to the current flowing in it) and its natural resistance increases. This resistance decreases the amount of current reaching the coil after the engine has warmed to its working temperature and permits more optimum coil design.

Some ballast resistors are encased in a ceramic form and mounted near the coil. In other cases the ballast is the wire that runs from the ignition switch to the coil. Many boats will have a system where the action of the starting solenoid or a special relay set of contacts will short out the ballast resistor during starting so as to allow full voltage and current to be supplied to the coil. After the engine starts the ballast is automatically inserted back into the circuit where it begins to function as designed.

Ballast Problems

When a ballast resistor is burned out and therefore an open circuit, the engine may start out but will die the minute the key is

Fig. 7.2 Example of a ballast resistor. Test with an ohmmeter if you suspect it to be faulty.

released to the run position. The reason, of course, is that the ballast is shorted out during starting but becomes an open circuit when the starter key is placed in the run position. Thus no current can get to the coil. Corroded terminals and increased resistance will also cause poor ignition performance. Ballast resistors cannot be repaired, but replacement is a simple task.

Coil

This device—the heart of the ignition system—is in reality a very high step-up ratio transformer, and in that sense calling it a coil is technically incorrect. Coil or transformer, it still works the same and does not normally require much maintenance. Basically, the coil is constructed of a few turns of copper wire in the primary and thousands of turns of fine magnet wire in the secondary. A typical ohmmeter reading of the resistance of the primary will show a very low (and normal) amount of resistance in the primary. The normal reading of resistance in the secondary will range from 8,000 to 10,000 ohms on most small test meters. If your boat uses a standard 12 volt battery, the voltage applied to the primary by way of the point set and ignition switch will be very close to 12 volts. As the cam shaft in the engine and therefore the distributor shaft rotates, the point set (switch)

Fig. 7.3 This plastic cased high voltage coil, otherwise excellent for boats, is electronically noisy.

closes and the contact points "dwell" together for a long period of time, allowing current to flow in the primary and thus a very strong magnetic field to build up and completely surround the secondary. As the cam in the distributor approaches a shoulder or high point, the point set is snapped open and the circuit to the primary of the coil is broken, which stops the flow of primary current. (The field suddenly collapses and magnetic lines of force cut across the thousands of turns of fine wire in the secondary.) When this happens, an electrical phenomenon occurs that causes up to 14,000 volts to be induced into the secondary wiring. This effect only lasts for a few thousandths of a second, but long enough for the voltage to be conducted to the distributor rotor, which is rotating in proper sequence to connect the high voltage to the appropriate spark plug where it will cause a bright blue spark to jump across the anode and base of the spark plug and ignite the fuel and air mixture within the cylinder.

Coil Troubles

Since there are no moving parts within the coil to wear out, very little can go wrong. However, in damp salt air some or all of the high voltage at the coil tower can and often does leak to a nearby ground. Thus the proper amount of high voltage insulation must be maintained or the plugs will fire poorly or not at all.

The coil usually is mounted directly upon and grounded to the engine. Thus the mounting is exposed to corrosion problems and vibration when the engine is running. PPM must include a regular check of the coil mounting hardware for both cleanliness and tightness. The low voltage connections (small threaded studs) found on the insulated perimeter of the coil are also subjected to rust and corrosion and must be kept clean and insulated. Low voltage wires must be attached to these studs using soldered crimp terminal lugs with complete rings—not the fork type. A lock washer under the retaining nut is a must so that the terminals and lugs do not work loose under vibration. Spray periodically with an electrical insulating spray such as WD-40 or CSC.

In your PPM program pay particular attention to the high voltage tower. Remove the high voltage wire from the coil tower to the distributor and carefully inspect the lead itself looking for signs of green-grey corrosion on any of the metal tips. The metal connector is down inside of the insulated coil tower and must be inspected with a small flashlight to make sure that the surface of the metal is clean and shiny. If you do not have a metal coil tower brush, use the eraser end of a pencil wrapped with a fine sandpaper to clean the metal and then blow out any left-over dirt. Spray with insulation protection. Carefully examine the insulated tower for any signs of cracks or carbon tracks that will hide a crack. A cracked insulator will cause high voltage leaks and is not repairable. The coil must be replaced. From time to time spray the entire top of the coil, and that alone will prevent most high voltage problems.

Noise caused by coils, high voltage spark plug wires, and spark plugs is covered completely in Book 4 of this series, *Electronic Equipment Care and Repair*.

Ignition Wires

Wires or cables can and often do cause electrical failures. Moisture, salt, and corrosion in the presence of very high voltage cause periodic breakdowns in the insulation resistance of the wires. A PPM skipper will keep on hand an ignition wire that is equal in length to the longest spark plug lead on the engine and that has one terminal affixed at the spark plug end so that the coil end can be attached in the event a lead fails.

Fig. 7.4 Choose ignition wire carefully; use only the best.

The ignition "harness" is usually a heavily insulated wire where the voltage is conducted through a single center conductor of carbon impregnated string or a resistance alloy. This type of wire causes less interference to electronic equipment and provides for a better spark to the plug. PPM requires only that the wire be carefully inspected each season for chafes, wear, and cracks in the insulation. Be especially suspicious of the rubber boots that cover the spark plug ends for they get very hot and often break down. Make sure that spark plug leads are carefully dressed with no crossing of wires close to each other. See that each plug wire is held tightly in its own place on the plug wire looms and that the insulating rubbing bushings in the looms are in good condition.

Distributor

Most small craft skippers seem to view the distributor as an overly complex device that must not be touched except by experts, but it is no more than a couple of electrical-mechanical switches. At the top of the distributor there is the distributor cap, a troublemaker if there ever was one! The distributor cap is a molded plastic device with a small smooth metal button at the center of the inside. As the rotor spring terminal presses on this button, it conducts the high voltage from the coil to the end of the rotor. Small metal tabs are also found inside the base of the distributor cap at each of the spark plug towers. Because of heat, vibration, extremely high voltages, and moisture, the distributor cap will often develop minute cracks inside, where they are missed by all but the most careful inspections. These cracks cause hard or no starting and bad misfiring of the plugs. The surface of the cap will accumulate salt crystals and dirt which will short out the high voltage if not kept spotlessly clean, dry, and insulated by the PPM skipper.

Rotor

This small insulated device is not normally a troublemaker and is easily and inexpensively replaced. It is a good idea to replace the rotor every other season and put the old rotor (after cleaning and burnishing the metal contacts) in the possibles box. Thus a spare rotor is always available, if only to give to a non-PPM skipper in an emergency!

Point Set

Some new boats are equipped with breakerless ignition systems in which the point set has been eliminated. Points, which are no

Fig. 7.5 A distributor that has been removed from the engine to show layout of parts.

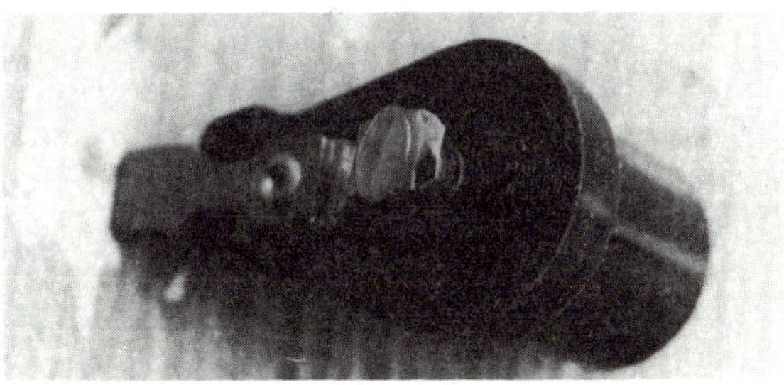

Fig. 7.6 Example of a replacement rotor for the distributor.

Fig. 7.7 Replacement set of points and its condenser.

more than switch contacts, are traditionally troublemakers. A PPM skipper knows that most of the things that go wrong with points are the result of normal wear and takes steps to anticipate problems by either adjusting the points at appropriate intervals or replacing, adjusting, and timing the engine each season.

Point troubles stem from such things as normal rubbing block wear. The fiber-composition rubbing block (or cam follower) rides on the distributor cam shaft and even when properly lubricated will wear to a certain extent. When this block wears, the dwell is increased, which has the effect of knocking the engine slightly out of time and decreasing engine performance in a number of ways. A quick check with a dwell meter will confirm or deny excessive rubbing block wear. A dwell meter is a good investment for the PPM skipper and its cost will be quickly recovered since it can be used on the family car as well as the boat.

A second source of point trouble is the faces of the points themselves, which when burned, out of alignment, or pitted increase the resistance to current flow to the primary of the coil. The effects of overall engine performance are very evident: misfiring, hard starting, excessive fuel consumption, and general poor performance at high RPMs. Most dwell meters have a point resistance check feature using a transfer function switch. If a dwell meter is not available, either an ohmmeter or DC voltmeter can be used in its place. If an ohmmeter is used, point resistance should not exceed a fraction of one ohm. If the voltmeter is used, accept no more than seven-tenths of a voltage drop across the points when the points are closed.

The point set is spring loaded, and very often the spring is ignored as a source of point problems. If it becomes weak through normal metal fatigue, the point set will bounce at high RPMs and cause missing and surges of power when you least want them. It is possible, though seldom practical, to adjust the spring point tension, if the tension is specified in the engine maker's manual, by bending it carefully with needle nosed pliers or by adding shims or washers to the mounting under the spring. It would in most cases be far better to simply install the spare point set that the PPM skipper always carries aboard.

The point set is mounted on the contact plate with two screws that in most cases are adjusted by means of an eccentric screw which sets the spacing of the point set.

On some marine eight-cylinder engines, there is a double point set, which has the advantage of being able to provide the coil with greater degrees of dwell. A double point set is installed and adjusted much the same as a single point set.

In summary, bad points will cause misfiring if the faces are corroded. Too much clearance (low dwell) causes engine knocking. Excessive dwell (low clearance) will cause backfiring through the carburetor and make the engine hard to start. Weak springs will cause high speed missing, and pitted points cause stalls. Broken points will kill the engine entirely.

Condenser

The condenser or a capacitor is connected electrically across the point set. To this day many otherwise sophisticated mechanics believe that the function of the condenser is to prevent arcing across the points. It does help to prevent point arcing, but its real purpose is to allow the coil to discharge its energy more rapidly by an electrical effect called "ringing" by some technicians. It rarely causes trouble, but it is customary, easy and inexpensive to replace the condenser each time the points are changed.

Fig. 7.8 Replacement condenser—be sure it is the proper value!

Lubrication

At the top of the distributor shaft you will usually see a small hollow space in which there is an oil wick. One or two drops of oil once a season is all that is needed, and oil dripping from the dipstick can be used. *Do not over oil*—you might cause burning and pitting of the points.

On some of the older engines a tiny dab of special grease was applied to the face of the cam to prevent excessive rubbing block wear on the cam follower. In the newer engines you will find a doughnut-shaped wick of plastic sponge that can be rotated to present a new face to the cam. On other distributors the wick may be a square-shaped bar that can be turned end for end to get another season out of it. Do *not* attempt to add grease to an old wick—simply replace it.

Spark Advance Mechanisms

Some distributors have two spark advance systems, the mechanical or centrifugal advance and the standard automotive vacuum advance. The value of a vacuum advance on a marine engine is questionable as its purpose is to provide spark advance for sudden accelerations and shifting of gears such as a car requires in traffic.

Fig. 7.9 End view of the distributor shaft showing the lubrication wick.

The mechanical advance system is located in the lower section of the distributor housing and is normally covered by the breaker mounting plate. Access to inspect or repair is gained by removing (generally) two breaker plate mounting screws. From underneath the mechanism can be viewed as two movable weights loaded with two springs. This system governs the engine speed with changes in load. As the engine accelerates, the weights tend to fly out and away from the distributor cam shaft and act to shift the position of the cam lobes in relation to the engine crankshaft. Thus, at high speeds the point set will break earlier in the cycle, and at low speeds the springs pull the weights in, retarding the spark for easy starting and smooth idling.

Few troubles are found in the marine engine except that the weights and springs will rust and tend to stick in one position. This cause of poor engine performance may be overlooked by even the best mechanics. The PPM skipper inspects the advance mechanism before he installs the new point set at the beginning of the boating season.

The vacuum advance system operates from the vacuum in the intake manifold. A rubber hose normally connects the vacuum advance to a small outlet on the manifold. Troubles here are

Fig. 7.10 The point set mounting plate has been removed to expose the mechanical advance weights.

usually traced to broken or punctured diaphragms within the unit. A broken diaphragm will bleed air back into the manifold and cause rough idling. The only cure is to either disconnect and blank off the manifold connection or renew the diaphragm if the unit is not sealed. Otherwise it must be replaced.

Spark Plugs

While not one of the persistent troublemakers in an internal combustion engine, the spark plug does from time to time get sick. Fortunately, it is a very simple device. It is easy to adjust the gap properly with the correct type of gauge. It is easy to remove and replace a plug.

Large horsepower outboard engines often have surface gap plugs, which take advantage of the higher spark potentials of modern ignition systems, such as the breakerless and capacitor-discharge types. The surface gap plug does not have adjustable gap points and will not perform well on conventional ignition systems.

The resistor type spark plug has become popular in small craft because of its low electronic noise characteristics. Most PPM skippers would do well to replace the plugs at the beginning of each season and carry not less than two spare plugs in the possibles box.

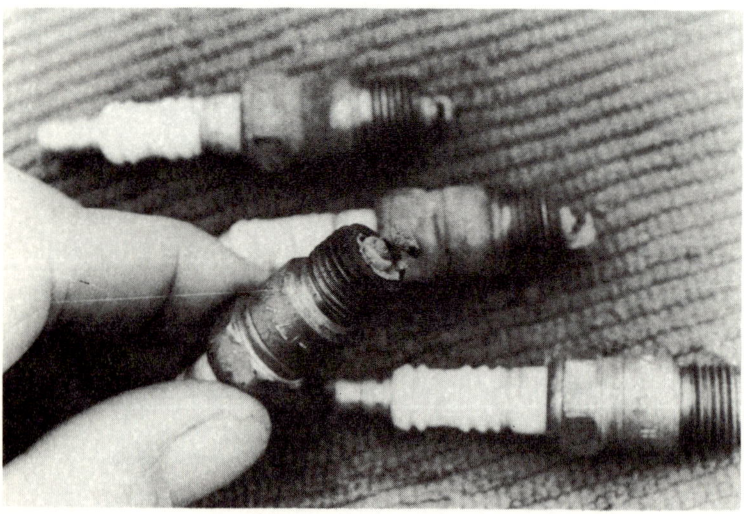

Fig. 7.11 Once the spark plugs have been removed, take the time to "read" them. This plug shows excessive ash deposits from poor grade of fuel.

What is not generally known about spark plugs (although there has been a substantial amount of publicity by the makers) is that a set of used plugs can be "read" to determine what is going on inside the engine. The condition and appearance of the business end of the plug often indicates an improper adjustment in the air-fuel mixture, impurities in the fuel or poor fuel, leaking piston rings, or other things that the PPM skipper needs to know about his propulsion system. Illustrated charts published by the spark plug makers show examples, often in color, of things to look for and interpret what might be wrong. Each time you change the plugs you would do well to read the used plugs.

Refer to your owner's manual for the recommended types of plugs for your engine; then set the specified gap with the proper gapping tool.

TUNING AN INBOARD GASOLINE ENGINE

Tools and Instruments

Gather the necessary technical data from the engine manual, such as spark plug type and gap spacing, point set spacing and dwell in degrees, and timing mark settings. Then assemble the following tools and test instruments:

Wrenches You will need at least a spark plug wrench. A deep throated socket wrench with rubber inserts for holding the plug will be helpful. If you can arrange to borrow or buy a torque wrench to set up the plugs according to specifications, so much the better.

Gauges A set of combination spark plug gauges and flat feeler gauges will be required to set the gap on a new set of plugs.

Test instruments Borrow or rent from an agency a timing light (avoid the neon type that cannot be seen in daylight) and combination dwell and resistance meter. If you purchase nonprofessional grade instruments, they will pay for themselves in the second tune-up that you do yourself. These inexpensive instruments differ from the professional only in the quality of construction. Professional instruments are made for rugged everyday use; the less expensive ones are intended to be used only two or three times a year and should last a lifetime with proper care.

A point set and matching condenser will also be needed. A can of insulating spray such as WD-40 or CSC-666 will be most helpful in preventing future problems.

Step-by-step Procedure

1. Loosen but do not remove the plugs until you have blown out the surrounding area with a strong air stream. If a compressed air hose is not available, a bicycle pump or a reversed vacuum cleaner will do nicely. There is even canned compressed air available if you cannot find any of the above. It is very important that dirt and carbon not be allowed to fall into the combustion chamber.

2. Remove and *read* all of the old spark plugs. If you have an older engine, check the compression parameters of each cylinder with a compression gauge. The compression test is simple and extremely revealing as to the overall internal condition of the engine. (The detailed step-by-step instructions for this test are given in Book 2 of this series.)

3. Set the gap recommended in the owner's manual for your spark plugs and apply a drop or two of oil to the threads before you start them back into the threaded holes in the engine block. Start each plug back in the hole *by hand*! It is essential that the plug not be driven into the block crossthreaded. If the engine is hot, you can use a small length of rubber hose to hold the plug as you start it into the hole. Torque the plugs up tightly to specification. If too tight, they may freeze in the block; if too loose, they can cause blow-by, which will burn out the plug and could start a fire.

4. Loosen and remove the distributor cap without removing any of the spark plug wires. Generally there will be enough slack in the wires to allow the cap to be dressed back out of the way. Using a strong light, carefully inspect the *inside* of the cap for cracks and carbon tracks. Replace the cap on the least suspicion. If you do replace the cap, orient the new cap with the old one and replace the spark plug wires one by one so that you do not get them crossed. Now is a good time to wash the cap inside and out with a degreasing solvent or alcohol. After it dries give it a good spraying with insulating spray.

5. Remove the rotor. Inspect and burnish clean the metal section. It is a good idea to install a new rotor (they are not expensive) and keep the old cleaned-up one in the boat box as a spare.

6. With the rotor removed the point set and condenser will be exposed. Take your time to study how the point set is mounted on the point set and condenser plate. Use the new point set as a

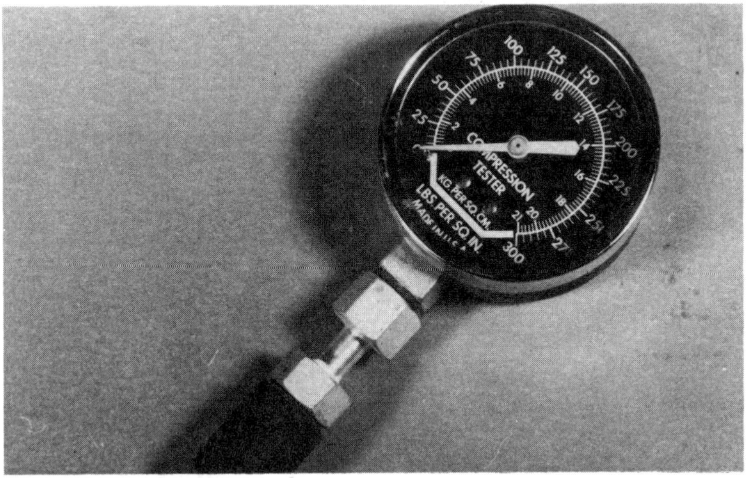

Fig. 7.12 When all of the spark plugs are out of the engine, take time to do the compression test.

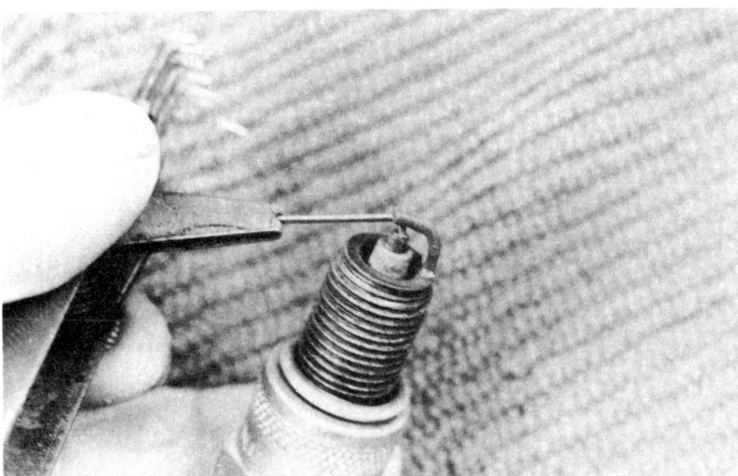

Fig. 7.13 Set spark plug gap with proper round space gage.

reference. Note where the distributor wire is attached to the point set—condenser lead will usually be found attached at the same point. When you are satisfied, loosen and remove the fixed plate screw. Loosen the second screw so that the point set can slide under it. (It is usually not necessary to remove this

Fig. 7.14 Loosen and lift up the distributor cap, but do *not* take off the high voltage plug wires. Dress the cap back out of the way of your work.

Fig. 7.15 Remove and inspect the rotor; it may need to be cleaned or replaced.

Fig. 7.16 Remove the point set and condenser.

screw.) Note the position of the eccentric adjusting screw which is used to set point gap spacing. Remove and inspect the faces of the breaker points. Are they burned? Pitted? Worn on one section? These problems indicate a maladjustment in the installation. With the breaker points out of the distributor, loosen and remove the condenser clamp screw. Remove and discard the old condenser. It is not easy to determine its condition without elaborate electronic testing equipment, and it is best to carry a new one as a spare.

7. Study how the mounting plate is installed in the distributor body. Usually no more than two heavy screws and lock washers hold it in place. Remove the screws and washers carefully so they will not fall into the distributor body. Remove the plate and set it aside to expose the centrifugal advance weights. Note that some distributors have the advance weights mounted above the point set. Inspect the condition of the weights and springs making sure that there is no rust or corrosion present and that the weights are free to move. Spray with WD-40 and replace the point set mounting plate back into its position.

8. Now install the new point set and condenser but do not fully tighten the breaker point set mounting screw just yet.

9. Using a "bump" switch or any convenient method, move the engine crankshaft (and therefore the cam shaft) in the small-

Fig. 7.17 Remove the mounting plate (under finger) so that the advance weight may be inspected.

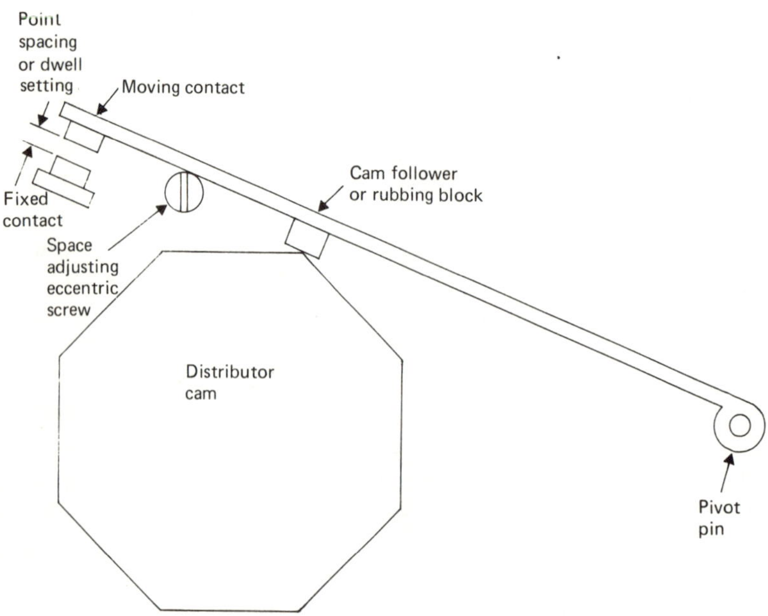

Fig. 7.18 Diagram of the point set spacing area.

est of steps until the rubbing block of the point set is resting on one of the cam shoulders at its highest point.

10. Select the size of flat feeler gauges as specified in the engine manual. Note the present gap of the point set. Use a screwdriver to open the gap so that the feeler gauge can be inserted into the gap. Readjust the gap by turning the eccentric screw with the screwdriver with the gauge in place. The gap is correct when the gauge can be slid back and forth with just a slight drag on it. If the gap is a tiny bit oversized, it will allow for rubbing block cam follower wear, which usually takes place immediately after installation and then settles down. Two-thousandths of an inch oversize is said to be about the right amount. In any event the dwell meter will tell you exactly where you stand for proper point gap spacing.

11. Inspect and turn the lubricating sponge (if used). If there is no sponge, lubricate the cam with a *tiny* amount of high quality grease. If there is an oil wick on the top of the cam shaft, give this a couple of drops of engine oil from the dipstick.

12. Replace the rotor and the distributor cap and give everything a final check for connections and tightness.

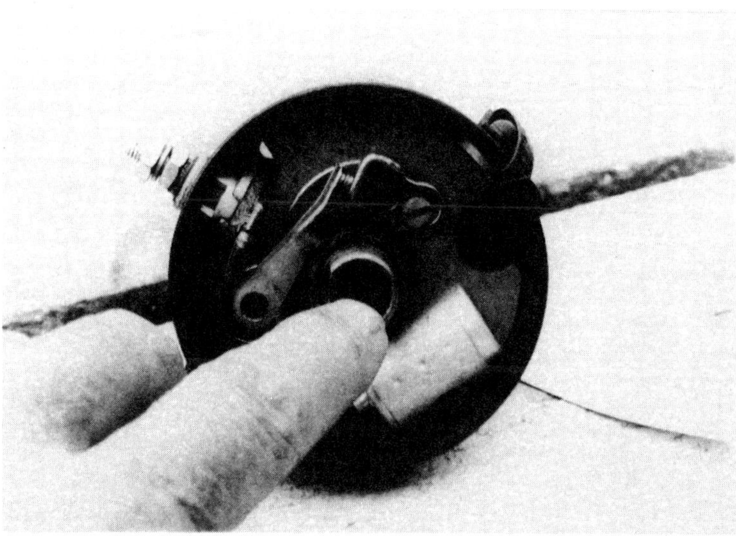

Fig. 7.19 Inspect the lube wicks (there may be more than one). Add a drop or two of oil from the dipstick if needed.

13. Study the instructions for the use of the dwell meter and begin with the point resistance check. This important check will tell you if the points are properly aligned when closed. If more resistance shows than is allowed, you will have to remove the cap and rotor and inspect the position and condition of the points when in the *closed* position.

14. Connect the dwell meter for the dwell test and start the engine. Be sure to set the dwell meter function switch to the dwell position and the selector switch for the number of cylinders in your engine. With the engine running, note the degrees of dwell shown on the meter and compare them with the specifications in the engine manual. If the dwell is too high, the point spacing is too small; and conversely, if too low the point spacing is too great. For either condition you will have to stop the engine and readjust the point gap again. Do not expect to get it exactly on to the precise degree. It is going to change very quickly as the rubbing block wears anyway.

15. With the dwell set to + 2° you can now shut down the engine and consult the instructions for timing light. Leave the dwell meter connected for the moment as it will be used for further tests. Connect the timing light according to the instructions. Normally, the timing light is connected to number one

Fig. 7.20 Study the instructions for the combination dwell/tachometer test meter. There are several tests that you can do.

spark plug (high voltage lead) and the two smaller leads are connected to the positive and negative terminals of the battery to provide power to the light.

17. From the engine manual you will have determined the specified number of degrees of advanced timing. Locate and identify the timing marks, which are normally found on the idler pulley on the front of the engine and a little plate attached to the body of the engine. It will often be necessary to bump the engine over until the timing mark shows on the idler pulley. Use a piece of chalk or crayon to outline the timing marks on both sides. Once you have located and emphasized the marks, restart the engine and focus the timing light on them. If the engine is in time, the marks will coincide. If the marks are not in coincidence, loosen the distributor clamp bolt at the base of the distributor so that the distributor body can be turned by hand to set the time. Again focus the light onto the marks and rotate the body of the distributor until the marks are in coincidence. Tighten the distributor clamp bolt making sure the distributor does not change position. Recheck the time. Essentially the engine is now partially tuned.

18. One last check will assure that the advance system is working properly. Again focus the timing light on the marks with the

Fig. 7.21 Engine timing marks are found near the balance wheel or idler pulley on the front of the engine.

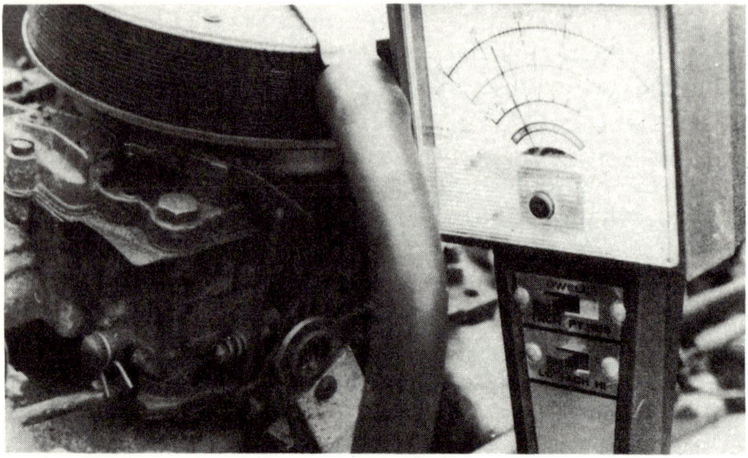

Fig. 7.22 Use the dwell/tachometer to set the mixture valves. Note screwdriver blade at lower left corner of photograph.

engine idling and manually advance the throttle as you note the movement of the marks. The instructions with the timing light will generally tell you what to look for.

19. If your dwell meter has a tachometer function (as most do), set the meter to read RPMs and set the idle adjust screw to the number of RPMs specified in the engine manual. The idle adjusting screw is usually a spring loaded screw on the carburetor and will be found close to the throttle cable ends.

20. For a complete tune-up carry out the air-to-fuel ratio test and reset the mixture screws on the base of the carburetor following the instructions in the dwell meter handbook. These instructions will also include a simple test that should be carried out to determine the cylinder balance.

An engine tune-up by the PPM skipper who follows all of the above steps will be more complete than one done by an average mechanic.

Since more than half the maintenance problems that beset the average small craft are traced to electrical problems the application of the suggestions and techniques in this book should eliminate 50 per cent of your problems. Don't be afraid to try - it's really easy. It is not dangerous as is working around house wiring and AC voltage. It pays off in a better and longer running boat.

INDEX